AIRBORNE
OCCUPATIONAL
HAZARDS
IN SEWER SYSTEMS

AIRBORNE
OCCUPATIONAL
HAZARDS
IN SEWER SYSTEMS

AMY FORSGREN
KRISTINA BRINCK

CRC Press
Taylor & Francis Group
Boca Raton London New York

CRC Press is an imprint of the
Taylor & Francis Group, an **informa** business

Cover design by Erik D. Forsgren

CRC Press
Taylor & Francis Group
6000 Broken Sound Parkway NW, Suite 300
Boca Raton, FL 33487-2742

First issued in paperback 2019

© 2017 by Taylor & Francis Group, LLC
CRC Press is an imprint of Taylor & Francis Group, an Informa business

No claim to original U.S. Government works

ISBN-13: 978-0-4987-5787-4 (hbk)
ISBN-13: 978-0-367-87680-7 (pbk)

Library of Congress Cataloging-in-Publication Data

Names: Forsgren, Amy, author. | Brinck, Kristina, author.
Title: Airborne occupational hazards in sewer systems / Amy Forsgren and Kristina Brinck.
Description: Boca Raton : Taylor & Francis, CRC Press, 2017. | Includes bibliographical references and index.
Identifiers: LCCN 2016014831 | ISBN 9781498757874
Subjects: LCSH: Sewerage--Safety measures. | Manholes--Safety measures. | Sewer gas--Toxicology.
Classification: LCC TD653 .F67 2017 | DDC 628/.20289--dc23
LC record available at https://lccn.loc.gov/2016014831

Visit the Taylor & Francis Web site at
http://www.taylorandfrancis.com

and the CRC Press Web site at
http://www.crcpress.com

To Kavi, who demanded more pictures.

Contents

Foreword by Markus Holmberg

Our vision is zero fatalities in the industry. Awareness of hazards is key, together with appropriate personal protective equipment and safe working methods.

There is never enough knowledge to completely eliminate risk. We have to manage risk: with good risk analysis processes and commitment from all levels of management we can reduce hazards. We must always strive to provide the people in our industry with as much protection as possible, through safe product design, safe systems, good operating methods, and increased knowledge.

Knowledge of hazards goes hand in hand with safe designs and operating methods. Many in the industry are already very good at educating their workforce about the potential hazards in the wastewater field. But there is room for improvement, and this book is a step in the right direction.

Safe products, safe systems, good methods, and knowledge of the dangers: we cannot eliminate risk, but we can create awareness of it and thereby help manage it.

Markus Holmberg (Civ. Ing.)*

---

* Markus Holmberg has more than 18 years of experience as a technical expert on pump design and the wastewater industry. He currently serves as Manager of Documentation and Product Management for Xylem Inc., a global water technology company.

Foreword by Vikram Nanwani

The loss of a single human life in the workplace should be treated as unacceptable and as a rallying call for change and improvement. I believe it is the collective responsibility of industry, academia, and government to ensure we eliminate and, where this is not possible, manage all hazards found in all kinds of working environments, including sewer systems, to drive toward an environment of zero accidents.

Over the last decades we have seen huge improvements in this area throughout all industries and a significant change in the mindset toward these hazards. In most controlled environments, such as factories and construction sites where hazards are well known, predictable, and, in the majority of cases, visible, it is easier to manage or even eliminate hazards. However, in sewer systems it is much more variable, less controlled, and many of the key dangers are airborne—often with no warning signs before they become fatal.

Airborne Occupational Hazards in Sewer Systems brings together critically important technical knowledge and understanding of airborne dangers in sewer systems. This is a critical step in addressing this problem by providing practitioners with the awareness required to ensure the required actions are taken to protect those working in these environments. I believe that this book is an important step for the wastewater community in moving toward what should always be the number one goal for any industry—zero accidents and fatalities.

Vikram Nanwani (CEng FIMechE)*

---

* Vikram Nanwani has more than a decade of experience in the water and wastewater industry. He currently serves as Technology Director for Xylem Inc., a global water technology company.

Acknowledgments

Many people helped make this book possible. In particular, we thank (in alphabetical order)

- William C. Barrow, of the Michael Schwartz Library at Cleveland State University, for aid in Chapter 8
- Erik D. Forsgren, for the book's cover design
- Dr. Per-Ola Forsgren, of PerkinElmer, Inc., for discussions on biomarkers and hepatitis
- Prof. John Grabowski of Case Western Reserve University, for discussions about Chapter 8 and for putting us in touch with the Western Reserve Historical Society archives
- Dr. Per Hedmark, application development engineer at Xylem Inc., for discussions on hydrogen sulfide
- Magnus Ångman, Wikipedia editor on World War II, for aid with the introduction to Chapter 8

About the Authors

Amy Forsgren received her chemical engineering education at the University of Cincinnati (Cincinnati, Ohio) in 1986. She then did research in coatings for the Swedish paper industry for three years, before moving to Detroit in the United States. There she spent six years in anticorrosion coatings research at Ford Motor Company, before returning to Sweden in 1996 to lead the protective coatings program at the Swedish Corrosion Institute. Since 2008, she has been working in the water and wastewater industry at Xylem Inc., a global water technology company. Amy Forsgren lives in Stockholm with her family.

Kristina Brinck is information designer at Xylem Inc., Rye Brook, New York, where she has been working since 2005. When she received her diploma in Swedish Language Consultancy from Stockholm University back in 1998, Kristina had already been working with publications for 10 years, as journalist and translator. She then worked as Information Mapping® instructor and language consultant for five years before she started working at Xylem. Kristina's greatest area of interest is in the translation of information from one discourse to another, for one profession to understand another, or for complex information to make sense for a wide range of audiences.

Acronyms

ACGIH	American Conference of Governmental Industrial Hygienists
AGA	American Gas Association
AIDS	Acquired immunodeficiency syndrome
AM	Alveolar macrophage
Anti-HAV	Antibodies to hepatitis A virus
Anti-HEV	Antibodies to hepatitis E virus
ASTM	American Society for Testing and Materials
ATP	Adenosine triphosphate
ATSDR	Agency for Toxic Substances and Disease Registry (United States)
A/V ratio	Biofilm area/liquid volume ratio
BGS	British Geological Survey
BMA	Bone marrow aspirate
Btu	British thermal unit
CAP	Compound action potential
CAS	Chemical Abstracts Service
CCOHS	Canadian Centre for Occupational Health and Safety
CDC	Centers for Disease Control and Prevention (United States)
CFK	Coburn–Forster–Kane model
CFOI	Census of Fatal Occupational Injuries (U.S. Bureau of Labor Statistics)
CNS	Central nervous system
CO	Carbon monoxide
COD	Chemical oxygen demand
COHb	Carboxyhemoglobin
CT	Computed tomography
CUPE	Canadian Union of Public Employees
DHHS	Department of Health and Human Services (United States)
DO	Dissolved oxygen
ECG	Electrocardiograph
ELISA	Enzyme-linked immunosorbent assay
EOG Co.	East Ohio Gas Company
EPA	See "U.S. EPA"
EU	European Union
EU/m^3	Endotoxin unit per cubic meter
FDA	Food and Drug Administration (United States)

FEV1	Forced expiratory volume in 1 second (spirometry parameter)
FEV1/FVC	Ratio of FEV1 to FVC (spirometry parameter)
FVC	Forced vital capacity (spirometry parameter)
GBV	Hepatitis G virus
GNB	Gram-negative bacteria
GRP	Glass-fiber reinforced plastic
HAV	Hepatitis A virus
Hb	Hemoglobin
HBO	Hyperbaric oxygen
HBV	Hepatitis B virus
HCV	Hepatitis C virus
HDV	Hepatitis D virus
HEV	Hepatitis E virus
HFMD	Hand, foot, and mouth disease
HGV	Hepatitis G virus
HIV	Human immunodeficiency virus
HRT	Hydraulic retention time
HSE	Health and Safety Executive (United Kingdom)
IARC	International Agency for Research on Cancer
IDLH	Immediately dangerous to life or health (concentrations)
IGS	Institute of Geological Sciences (United Kingdom)
IHA	Indirect hemagglutination assay
J (e.g., Roberts J)	High Court Judge (England and Wales)
LEL	Lower explosive limit
LFL	Lower flammable limit
LJ (e.g., Roberts LJ)	Lady/Lord Justice of Appeal (England and Wales)
LNG	Liquefied natural gas
LOC	Limiting oxygen concentration
LS&R	Liquefaction, storage, and regasification
LTEL	Long-term exposure limit (United Kingdom)
MA	Methanogenic archaea
MAT	Microscopic agglutination test
MOC	Minimum oxygen concentration
MRI	Magnetic resonance imaging
NFPA	National Fire Protection Association
NIOSH	National Institute for Occupational Safety and Health (United States)
NO	Nitrous oxide
NOHSC	National Occupational Health and Safety Commission (Australia)
NRC	National Research Council
NWWA	North West Water Authority (United Kingdom)
OELV	Occupational exposure limit value (EU)

OSHA	Occupational Safety and Health Administration (United States)
PCF	Pericardial fluid
PCR	Polymerase chain reaction
PDM	Pittsburgh-Des Moines Steel Co.
PE	Population equivalent
PEL	Permissible exposure limit (United States)
ppb	Parts per billion
PPE	Personal protective equipment
ppm	Parts per million
PVC	Polyvinyl chloride
qPCR	Real-time polymerase chain reaction
RADS	Reactive airways dysfunction syndrome
REL	Recommended exposure limit (United States)
SCBA	Self-contained breathing apparatus
SCOEL	Scientific Committee on Occupation Exposure Limits (EU)
SLOD	Significant Likelihood of Death (HSE, United Kingdom)
SRB	Sulfate-reducing bacteria
STEL	Short-term exposure limit
STEL-C	Short-term exposure limit—ceiling
STP	Standard temperature and pressure
TLV®	Threshold Limit Value
TWA	Time-weighted average
UEL	Upper explosive limit
UFL	Upper flammable limit
U.S. EPA	U.S. Environmental Protection Agency
v/v	Volume percent
WEL	Workplace exposure limit (United Kingdom)
WHO	World Health Organization
WHS	Work Health and Safety Act (Australia)
w/w	Weight percent
WWTP	Wastewater treatment plant

1

Introduction

1.1 Why a Book on Airborne Occupational Hazards in Sewers?

Sewers are dangerous places to be, because

1. They are confined spaces
2. In these particular confined spaces, there is always the possibility of insufficient oxygen, toxic gases, explosive gases, and infectious agents

Sewer work is carried out in cramped spaces with little or no ventilation, little room for movement, and few exits or entrances.

Sewers are designed to convey a substance—human wastewater—with high organic load, undergoing constant biological decay. The breakdown of this organic matter produces a witch's brew of toxic gases (e.g., hydrogen sulfide) and explosive gases (e.g., methane) and can consume some or all of the atmospheric oxygen. The organic matter is also extremely rich in microbes, both dead and alive. Some of them are disease-causing pathogens that can survive, even flourish, in the conditions found in the sewer system.

There are books and journal articles available on the occupational hazards of wastewater treatment plants. Consistently (and irritatingly), these publications say, "…but this particular hazard is minimal, unless you have to enter a sewer" or "This presents a hazard only in the collection system…." We decided that it was time to give center stage to the occupational hazards sewer workers face.

These hazards cannot be eliminated, since microorganisms and biological decay are an inherent part of sewage. But intelligent precautions can do a great deal to minimize the exposure to dangers. We hope to raise awareness of the hazards, so that the precautions are taken.

1.2 Scope of the Work

This book provides technical information about the toxic or explosive gases, and infectious agents, most commonly found in collection systems.

The book does not provide information on legal requirements and regulations. The reader is urged to consult legal experts for questions about applicable laws and regulations. Nor does this book provide medical advice: the information used here should not be used for self-diagnosis or to determine treatments.

Many of the dangers that may arise in a collection system are very well covered elsewhere. For that reason, we have left out hazards that are common to many industries, such as

- Electrical hazards
- Slips, trips, and falls
- Machinery

We have concentrated on those hazards that are not so common in other industries and therefore are not covered as thoroughly as we would like.

1.2.1 Confined Spaces

As stated at the beginning of this chapter, sewers are confined spaces. We begin accordingly by giving a brief overview of confined spaces in Chapter 2. Many readers will already have extensive knowledge of confined space hazards, precautions, and regulations; they may wish to skip Chapter 2 and go directly to Chapter 3.

1.2.2 Most Common Airborne Hazards

The toxic or flammable gases upon which we have concentrated are hydrogen sulfide and methane. H_2S receives three chapters in this book; methane rates one chapter plus two case studies. These two substances are produced by the biological decay of sewage and can therefore be expected to exist in relatively large quantities in collection systems.

Carbon monoxide, carbon dioxide, oxygen deficiency, ammonia, and gasoline are also covered; but these are common industrial hazards, not at all unique to, or characteristic of, our industry. They are covered often, and well, in other safety books.

The chapters dealing with biological hazards are focused on bacterial, viral, and fungal and parasitic agents most commonly found in sewage that cause infectious diseases. Parasites, though perhaps not really an airborne hazard, are briefly touched on, simply because it seemed odd to exclude them entirely from the biological review.

1.3 Some Useful Background Information

Several of the chapters in this book deal with flammable/explosive gases and make repeated reference to lower flammability limit (LFL), lower explosive limit (LEL), limiting oxygen concentration (LOC), and so on. A brief explanation is provided in this section of what the terms mean as used in this book.

Also, the North American readers will be familiar with the National Institute for Occupational Safety and Health (NIOSH), Occupational Safety and Health Administration (OSHA), American Conference of Governmental Industrial Hygienists (ACGIH), and so on; but for readers from other continents, these terms are briefly explained in Section 1.3.2.

1.3.1 Explanation of LEL/UEL, LFL/UFL, and LOC

The LFL is the minimum concentration of fuel in air for combustion to occur; below this the fuel/air mixture is too "lean." The upper flammability limit (UFL) is the maximum concentration of fuel in air for combustion to occur; above this the fuel/air mixture is too "rich." These are also known as lower explosive limit and upper explosive limit (LEL and UEL).

The LOC* is the minimum concentration of oxygen for combustion to take place; below this amount, combustion is not possible, no matter what the fuel concentration is.

It is important to note that these properties do not define hard-and-fast boundaries between "safe" and "unsafe" conditions. They are not fundamental properties of each chemical species. Instead, they are the result of measuring particular gas mixtures in particular apparatuses using particular methods [1]. They can be expected to change with circumstance, for example, when adding more gases to the mix.

1.3.2 Who's Who and What's What for Occupational Exposure

The ACGIH is a well-respected nongovernmental organization whose recommendations generally carry a lot of weight. The ACGIH develops and publishes guidelines called Threshold Limit Values (TLVs®) for exposure to workplace chemicals. TLVs are the airborne concentration of a substance that nearly all workers may be exposed to repeatedly without adverse health effects [2].

The ACGIH TLVs are frequently incorporated into legal regulations governing workplace chemical exposure. They may also be incorporated into industry standards issued by associations such as the National Fire Protection Association.

The NIOSH is part of the U.S. federal government (Department of Health and Human Services). NIOSH tests equipment, evaluates and approves

---

* Also known as MOC, minimum oxygen concentration.

TABLE 1.1

Comparable Terms

Terms in United States and Canada	Terms in Great Britain
PEL	WEL, workplace exposure limit (EU: OELV)
TWA	LTEL, long-term exposure limit
STEL	STEL

respirators, conducts studies of workplace hazards, and proposes standards to OSHA. NIOSH scientists determine the concentration that should be regarded as immediately dangerous to life or health (IDLH concentration) and a Recommended Exposure Limit (REL).

The European Union's analogous body to NIOSH is the Scientific Committee on Occupational Exposure Limits (SCOEL).

OSHA is the part of the U.S. federal government (Department of Labor) that adopts and enforces health and safety standards in the United States. OSHA uses the term Permissible Exposure Limit (PEL) to define the maximum concentration of a specific chemical to which an unprotected worker may be exposed. A specific chemical may have one or more types of PEL:

- TWA: Time-weighted average, usually set for an 8-hour workday
- STEL: A 15-minute short-term exposure limit that should not be exceeded during a workday
- C: Ceiling concentration that should not be exceeded at any point during a working period

Table 1.1 shows the comparable terms used by Canada, Great Britain, and the United States for PEL, TWA, and STEL.

References

1. Crowl, D. A. (2012). Minimize the risks of flammable materials. *Chemical Engineering Progress*, 108(4), 28–33.
2. ACGIH. (1991). *Documentation of the Threshold Limit Values and Biological Exposure Indices*, 6th ed. American Conference of Governmental Industrial Hygienists: Cincinnati, OH, Vol. II, pp. 786–788.

2

Confined Spaces

2.1 Introduction

Sewers are dangerous because they are confined spaces containing biologically active material.

In this book, the term "confined space" is used for an enclosed area or restricted space that

1. Is not designed for people to work in continuously
2. Has limited or restricted means of entry or exit
3. May be hazardous to any person entering it due to
   a. Its design, construction, or location
   b. Its atmosphere
   c. The materials or substances in it
   d. Any other conditions relating to it

Confined spaces are notorious for causing fatalities through oxygen deficiency, explosive gases such as methane, or toxic gases such as hydrogen sulfide.

In the cases of oxygen deficiency and toxic gases, it is not uncommon for both workers and their would-be rescuers to be killed. In fact, it is estimated almost half of all workers who die in confined spaces entered in an attempt to rescue other workers.

2.1.1 Permit-Required Confined Space

An enclosed area that meets the three criteria given earlier is a confined space. However, not all confined spaces present immediate health hazards. An attic crawlspace in a family home meets the criteria for confined spaces, but is not normally considered immediately dangerous to life and health.

The combined spaces that are dangerous are those that contain low oxygen, toxic or explosive/flammable gases, engulfment risks, or biological hazards. When one or more of these is present, then the situation changes—depending on local regulations, the area becomes a "permit-required confined space." A crawlspace in an attic is a confined space; a sewer or manure pit is a permit-required confined space.

The permit to enter a confined space should not be confused with a hot-work permit.

2.1.2 Examples

It is not possible to provide a comprehensive list of confined spaces in the wastewater industry; examples include tanks, diked areas, sewers, manholes, and valve chambers. Figure 2.1 shows some examples of confined spaces in the wastewater industry.

Open pits or tanks may not seem at first to be particularly confined. However, the atmosphere in and above open tanks and pits, especially on wind-free days, can easily pose dangers to workers. Even if the risk of a toxic or explosive atmosphere is extremely low, the difficulty in getting workers out in a hurry meets the criteria of an enclosed space. The need for rapid evacuation can arise, for example, from engulfment hazards due to the collapse of a retaining wall or a sudden flood of water into the system. Also, delay in getting an injured worker out of the pit risks aggravating or compromising his condition.

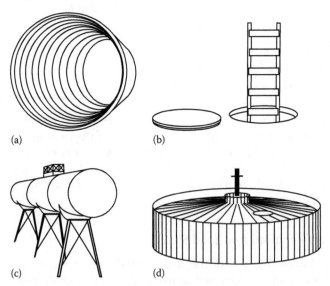

FIGURE 2.1
Examples of confined spaces in the wastewater industry. (a) Pipe. (b) Manhole. (c) Tank. (d) Digester. (From National Institute for Occupational Safety and Health (NIOSH), Preventing occupational fatalities in confined spaces, NIOSH Publication No. 86-110, U.S. Government Printing Office, Pittsburgh, PA, 1986.)

Sewers are hazardous confined spaces because workers can be engulfed and trapped in them when upstream water or sewage floods the system; toxic, flammable, or explosive gases may be present; pathogens may be present; and the air may be deficient in oxygen.

2.1.3 Misleading Name

The names "confined space" and "enclosed space" can be misleading, since they tend to conjure up images of chambers that are completely enclosed, that is, with floor, walls, and roof.

This is a false image, since open tanks or pits are also considered confined spaces. The walls create a space that is not freely ventilated, and there is a possibility of a hazardous atmosphere inside the space, which can endanger workers entering the pit or open tank—or even working above it.

Trenches dug to lay pipes or place manholes are also a type of confined space, with special problems [1]: cave-ins, falls, electrocution, being struck by heavy equipment or falling objects, and atmospheric hazards. Trenches near garbage landfills, for example, may be reasonably expected to contain a hazardous atmosphere [1,2]. And carbon monoxide, generated by blasting, has been known to migrate through soil into sewer excavations, with devastating results for those constructing the sewers.

Because they are open excavations, it is easy to overlook that trenches are confined spaces and to underestimate the dangers.

2.2 What Are the Dangers of Confined Spaces?

Confined spaces kill because

- They are unrecognized as confined spaces
- Most hazardous atmospheres are not visible to the human eye; many cannot be tasted or smelt at all
- The difficulty of exiting, before being affected by a hazard, is often underestimated
- Changing situations—Hazards can develop after entering the confined space, catching workers off guard

Figure 2.2 shows the relative importance of each hazard category.

The hazards presented by confined spaces can be divided into atmospheric, biological, and physical groups. Each of these is discussed in more detail in the following sections.

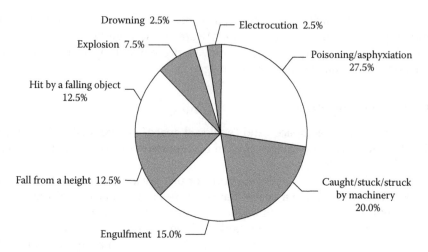

FIGURE 2.2
Confined space fatalities by category. (From Burlet-Vienney, D. et al., *Saf. Sci.*, 79, 19, 2015.)

2.2.1 Atmospheric

Atmospheric dangers—suffocation due to lack of oxygen, inhaling toxic gases, or ignition of flammable vapors or gases—are by far the greatest hazards in confined spaces.

2.2.1.1 Low Oxygen

Anoxia, also called hypoxia, is the inhalation of gases containing insufficient oxygen. Anoxia leads to pulmonary edema and petechial hemorrhages in the brain, lungs, and myocardium [3]. As the partial pressure of oxygen in the lungs' alveoli falls to 30–40 mm Hg or lower, the brain, organs, and tissues receive an insufficient amount of oxygen for normal metabolism. Loss of consciousness occurs rapidly. Someone entering an oxygen-deficient atmosphere with only 6%–10% oxygen will usually collapse within 40 seconds [4,5].

The oxygen concentration in normal breathing air is 20.9% [4,6]. Many jurisdictions statute a minimum oxygen level for performing work; in Alberta (Canada), for example, if oxygen falls below 19.5% by volume, then air-supplying respiratory protective equipment must be worn [7].

An oxygen-deficient atmosphere may arise from various causes:

- Other gases and vapors displace the oxygen.
- Chemical/biological processes such as the decomposition of organic matter by bacteria may use up oxygen.

- Combustion processes such as cutting or welding use up the oxygen.
- Groundwater can act on chalk and limestone to produce carbon dioxide and displace normal air [2].

Low oxygen levels cannot be detected by smell or sight; the air must be tested. A person who walks into an oxygen-deficient atmosphere is often completely unaware of the progression to unconsciousness. Once they have lost consciousness, death from asphyxia results unless oxygen is administered.

Table 2.1 and Figure 2.3 show the health effects associated with various levels of oxygen deficiency.

TABLE 2.1

Health Effects Associated with Various Percentages of Oxygen in the Atmosphere

% Oxygen in the Atmosphere	Health Effects
21 (normal oxygen content in air)	None
19.5	None
16	Impaired judgment and breathing
14	Faulty judgment, rapid fatigue
6	Difficulty breathing, death in minutes

Source: National Institute for Occupational Safety and Health (NIOSH), Preventing deaths of farm workers in manure pits, NIOSH Publication No. 90-103, U.S. Department of Health and Human Services, Centers for Disease Control, Morgantown, WV, 2007.

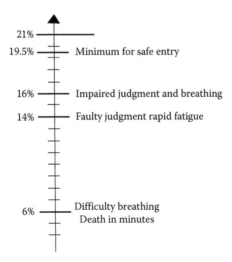

FIGURE 2.3
Oxygen scale. (From National Institute for Occupational Safety and Health (NIOSH), Preventing occupational fatalities in confined spaces, NIOSH Publication No. 86-110, U.S. Government Printing Office, Pittsburgh, PA, 1986.)

On September 15, 1986, oxygen deficiency in a sewer killed a plumbing contractor in Georgia (United States). Three workers were planning to connect a sewer line from a building to the main sewer line. In order to take measurements, one man entered the two-foot sewer manhole and climbed down 15 feet to the bottom, where he lost consciousness. The other two had remained outside the manhole, but seeing their coworker unconscious they entered the manhole in a rescue attempt. Before they could reach the victim, they both became dizzy and had to exit. After several failed rescue attempts, the fire department was called. Rescue squad personnel using self-contained breathing apparatus removed the victim and attempted resuscitation; at the hospital he was pronounced dead.

The cause of death was asphyxia due to oxygen deficiency. The atmosphere was tested after the victim was removed and was found to contain 20% methane and 6% oxygen and was negative for hydrogen sulfide and carbon dioxide.

Before entry no atmosphere tests were performed and the sewer vault was not ventilated. No confined space entry procedures were used; neither the plumbing contractor nor the larger contractor employing them was aware that a sewer manhole is a confined space and potentially hazardous to enter [8].

2.2.1.2 Flammable/Explosive or Toxic Gases

Because collection systems serve both households and industries in their communities, almost any chemical can, in theory, find its way into the sewer. It is beyond the scope of this book to list all of them; the principal flammable/explosive or toxic gases are shown in Table 2.2.

In Chapters 3 through 5, we will look more closely at hydrogen sulfide. Methane is examined in Chapter 6 and in the case studies in Chapters 7 and 8. Carbon monoxide, carbon dioxide, and other hazardous gases that might be found in sewer systems are described in Chapter 9.

TABLE 2.2

Flammability Properties, Vol.%

Chemical Species	Lower Flammability Limit (LFL)	Upper Flammability Limit (UFL)	Limiting Oxygen Concentration (LOC)	Toxic
Hydrogen sulfide	4.3	45.0	7.5	Yes
Methane	5.0	15.0	12	
Carbon monoxide	12.5	74.0	5.5	Yes
Gasoline	1.4	7.6	12	

2.2.2 Biological

Due to the nature of sewage, disease-causing bacteria (e.g., tetanus) and viruses (e.g., hepatitis), or parasitic microorganisms, may be present in the collection system. In Chapter 10 we examine biological hazards, and in Chapters 11 and 12 we discuss two specific diseases, hepatitis and leptospirosis.

2.2.3 Physical

The scope of this book is airborne hazards in collection systems. Physical hazards are well covered in other texts. Physical hazards in sewers can include

- Activation of machinery
- Heat
- Drowning (surfaces are often wet and slippery)
- Engulfment—for example, due to collapse of a wall
- Falling
- Falling objects
- Reduced visibility due to dust

2.3 Changing Conditions

If any changes occur during entry that may affect the safety of the workers, then workers have to be removed from the space until the new situation has been evaluated.

2.3.1 The Confined Space Changes Its Nature

Conditions inside the confined space can change suddenly and without warning. Hazardous material can be released into the space without warning; the carbon monoxide accident described in Section 9.1.5.2 is such an example. At the start of the workday, the atmosphere may have been safe; however, circumstances changed. Carbon monoxide from nearby blasting seeped through the soil and into the confined space, killing a worker [9].

Hazardous material can also be already present in a nongaseous state, and therefore not recognized—until, that is, it becomes vapor and suddenly creates an atmosphere that is immediately dangerous to life or health. Hydrogen sulfide dissolved in stagnant water is a classic example of this. Readings of the air above the liquid may not find anything alarming; but when the stagnant water is stirred up, dissolved H_2S becomes gaseous and creates an extremely hazardous atmosphere.

Hazards do not have to be dangerous gases, either; the sudden release of large volumes of water into a confined space has caused more than one fatality in the water and wastewater industry.

2.3.2 Changes in the Work to Be Done in the Confined Space

The nature of the work to be done in the confined space can also change. A typical example of this is entering a valve chamber for a preventive maintenance inspection. The initial plans are only to inspect the piping and fittings; however, the inspection may reveal corrosion, which makes it necessary to replace piping or fittings as soon as possible. If the corrosion is very advanced, it may not be possible to loosen bolted flanges; pipe cutting and welding (hot work) may be needed. The initial hazard evaluation of the space was based only on inspection of the equipment. When the need for hot work arises, a new hazard evaluation must be done.

2.4 Reducing the Risk

The best way to avoid confined space accidents is to never enter them; but this is very often not feasible. The second best way is to very strictly control *who* enters, and under what *circumstances*—in short, by establishing a program for confined spaces.

Key elements in reducing the risks of enclosed spaces include evaluating the hazards of the particular space, temporarily modifying the space, monitoring the air, having a means to rescue workers safely, and undergoing extensive training.

2.4.1 Confined Space Programs

In most localities, working in a confined space is strictly regulated. Aspects typically covered include the following:

- Which hazards must be evaluated?
- Who may perform the hazard assessment? (Note that in English-speaking countries, the terms "qualified person" and "competent person," when used in health and safety regulations, often have specific legal definitions.)
- What tests must be performed before entry?
- What must be monitored while workers are in the space?

- Under what conditions can workers enter the space?
- Power lockout/tag out.
- Mandatory personal protective equipment (PPE).
- Standby persons.
- Communication routines.
- Rescue procedures and equipment.
- Training of workers entering the space, standby persons, and rescue workers.
- Calibration and maintenance of test equipment, PPE, and rescue equipment.

All of this is usually documented in a written program. The exact contents of the program will depend on the regulations in force for the particular location. Occupational safety legislation, directives, and guidelines may vary from place to place.

2.4.2 Training

We cannot emphasize this enough: training workers in the dangers, and providing them with the right safety equipment and PPE, can be the difference between life and death when working in collection systems.

The minimum training needed is often specified in occupational safety legislation. The regulations in force depend on the particular locale.

Agencies such as the Occupational Safety and Health Administration (United States), Canadian Centre for Occupational Health and Safety (Canada), Healthy and Safety Executive (United Kingdom), and Safe Work Australia often provide excellent training information on confined spaces; a visit to their websites can be highly rewarding.

2.4.2.1 *Chain-Reaction Deaths Can Be Avoided*

"Chain-reaction" deaths are a well-known danger associated with confined spaces. Chain-reaction deaths are called thus because after the first victim is found in a confined space, a rescuer enters without proper precautions and equipment. The rescuer is in turn overcome, another rescuer enters and is likewise overcome, and so on [9]. Lee et al. [10] describe a scenario at a sewage plant that is all too common:

> A 24-year-old-male worker at a sewage disposal plant was transferred to the emergency medical center of a university hospital under the impression of myocarditis and acute myocardial infarction. He had suddenly collapsed two minutes after entering a manhole without respiratory protection to rescue an unconscious co-worker.

Rescuing coworkers who are trapped in confined spaces is challenging and dangerous. Estimates vary, but it seems clear that between one-quarter and one-half of all workers who die in confined spaces were trying to rescue other workers.

It is human nature to offer assistance to those in peril; but sadly, the untrained rescuer tends to die along with the trapped worker whom they were trying to save.

2.4.3 Testing the Atmosphere

Training the worker in the hazards is, in our opinion, the most important step to reducing the risk. The second step is to always know what atmosphere you're dealing with.

The atmosphere must be checked frequently with reliable, calibrated instruments. Continuous monitoring, with audible alarms, is even better. Local or national codes and regulations often stipulate what gases must be monitored and how often. Maximum levels are also codified; depending on the specific gas, there may be different levels for different exposure times. The maximum levels are also usually lower if multiple gases, with similar toxic actions, are present.

In general, the following need to be monitored [11]:

1. Oxygen-level indication—to determine the oxygen content of the air
2. Combustible gas vapor detection—explosive gases such as methane and vapors such as gasoline
3. Toxic gas detection—toxic gases such as H_2S, carbon monoxide, and nitrogen dioxide
4. Other air monitoring, as required—to detect other hazardous gases or vapors

A number of multifunction instruments are available that check for oxygen deficiency, flammable gases, and toxic gases.

It is necessary to be knowledgeable about the particular confined space, so that the correct gases are monitored. Monitors with sensors for chlorine and hydrogen sulfide will be of no use in a confined space that contains carbon monoxide and nitrogen dioxide [12]. Any toxic gases likely to be present will depend on the nature of the space (sewer? process tank?) and the work planned (painting, welding, etc.).

The sense of smell is useless for evaluating an atmosphere. Many dangerous gases have no odor at all [6]. It is impossible to detect an oxygen deficiency by sense of smell. And finally, hydrogen sulfide rapidly paralyzes the sense of smell; the higher the concentration of H_2S, the faster the loss of the sense of smell.

References

1. Occupational Safety and Health Administration. (2002). Excavations. Publication No. OSHA 2226. Department of Labor: Washington, DC.
2. Health and Safety Executive. (2013). Confined spaces: A brief guide to working safely. Publication No. INDG258. Health and Safety Executive: Liverpool, UK.
3. James, P. B. and Calder, I. M. (1991). Anoxic asphyxia—A cause of industrial fatalities: A review. *Journal of the Royal Society of Medicine*, 84(8), 493–495.
4. Williams, J. W. (2009). Physiological responses to oxygen and carbon dioxide in the breathing environment. In: *Presentation from NIOSH Public Meeting*, September 17, 2009, Pittsburgh, PA. National Institute for Occupational Safety and Health (NIOSH), Technology Research Branch: Pittsburgh, PA.
5. Appleton, J. D. (2011). User guide for the BGS methane and carbon dioxide from natural sources and coal mining dataset for Great Britain. BGS Open Report, OR/11/054. British Geological Survey: Keyworth, UK.
6. Gray, C., Vaughn, J. L., and Sanger, K. (2011). *Distribution/Collection Certification Study Guide*. Oklahoma State Department of Environmental Quality: Oklahoma City, OK.
7. Sewer Entry Guidelines. (2010). *Workplace Health and Safety Bulletin CH037*. Queen's Printer, Government of Alberta, Department of Employment and Immigration: Edmonton, Alberta, Canada.
8. Fatal Accident Circumstances and Epidemiology Database. (1986). Insufficient oxygen level in sewer claims the life of plumbing contractor in Georgia, incident FACE 8654. The National Institute for Occupational Safety and Health (NIOSH), U.S. Department of Health, Education and Welfare: Washington, DC. Accessed December 13, 2015.
9. Decker, J. A. et al. (1998). Department of Health and Human Services, Centers for Disease Control and Prevention, National Institute for Occupational Safety and Health. DHHS (NIOSH) Publication No. 98-122. U.S. Government Printing Office: Pittsburgh, PA.
10. Lee, E. C., Kwan, J., Leem, J. H., Park, S. G., Kim, H. C., Lee, D. H., Kim, J. H., and Kim, D. H. (2009). Hydrogen sulfide intoxication with dilated cardiomyopathy. *Journal of Occupational Health*, 51(6), 522–525.
11. Hydrogen Sulphide at the Work Site. (2010). *Workplace Health and Safety Bulletin CH029*. Queen's Printer, Government of Alberta, Department of Employment and Immigration: Edmonton, Alberta, Canada.
12. WorkSafeBC. (2013). Incorrect use of monitoring equipment in confined spaces can endanger workers. Bulletin WS 2009-03. WorkSafeBC Workers' Compensation Board of British Columbia: Vancouver, British Columbia, Canada.
13. National Institute for Occupational Safety and Health (NIOSH). (1986). Preventing occupational fatalities in confined spaces. NIOSH Publication No. 86-110. U.S. Government Printing Office: Pittsburgh, PA.
14. Burlet-Vienney, D., Chinniah, Y., Bahloul, A., and Roberge, B. Occupational safety during interventions in confined spaces. *Safety Science*, 79(2015), 19–28.
15. National Institute for Occupational Safety and Health (NIOSH). (2007). Preventing deaths of farm workers in manure pits. NIOSH Publication No. 90-103. U.S. Department of Health and Human Services, Centers for Disease Control: Morgantown, WV.

3

Hydrogen Sulfide, Part 1: The Macro View

3.1 Introduction

This occurs only too often in our industry:

> A male city sewer employee entered a sewer by ladder. He complained of
> a strong odor and started to climb out, then collapsed and fell back into
> the hole. The man remained in the hole for several minutes before the
> body could be retrieved. Men at the scene reported a strong hydrogen
> sulfide odor. [1]

Hydrogen sulfide, H_2S, is a deadly poison, as lethal as cyanide [2]. It can be
created by the decomposition of organic matter, such as wastewater transported in a sewer network.

H_2S is the second most common cause of fatal gas inhalation exposures
in the workplace, after carbon monoxide. H_2S accounts for 7.7% of such
cases [3–5].

It is highly toxic: 100 ppm is deemed immediately dangerous to life and
health. At 500 ppm—or sometimes lower—H_2S can cause unconsciousness
in a few seconds (the "knockdown" or "slaughterhouse sledgehammer"
effect). Death quickly follows if the victim is not immediately removed to
fresh air. At 1000 ppm, death can occur in a few breaths.

The toxicology of hydrogen sulfide is explored in detail in Chapter 4. In this
chapter, we present an overview of why it is such a problem in our industry.

3.1.1 Slaughterhouse Sledgehammer

It is impossible to exaggerate the speed with which H_2S overwhelms its victims. H_2S can cause an instant loss of consciousness; death from paralysis
of the respiratory center follows if the victim is not immediately removed to
fresh air. This is the "slaughterhouse sledgehammer" effect, also known as
the infamous "knockdown" of the sour gas and oil industry [2,6–11].

The loss of consciousness and collapse is so abrupt that witnesses often say
it was as if a power switch were turned off.

Knight and Presnell [12] have provided a graphic description of the speed with which this happens:

> Two gentlemen, aged 20 and 31 years, were surveying near 2 sewer man-holes in an industrial area of a suburban county near the coast of South Carolina. The younger man reportedly opened the manhole cover to enter it and suddenly yelled and fell in. The older man then ran over to the manhole to provide assistance and also suddenly fell in. Bystanders called EMS, and rescue personnel were equipped with SCBA respirators while retrieving the victims, who were said to appear blue. Both men were pronounced dead at the scene; no resuscitation was performed. County Water and Sanitation Authority employees reported an air con-centration of H_2S of 34 ppm, measured before the bodies were retrieved, but some time after the incident. Bystanders reported a strong smell of 'rotten eggs,' which dissipated over time. [12]

3.1.2 Secondary Victims

H_2S is so toxic that healthcare workers who treat victims of H_2S poisoning can themselves suffer ill effects. Ruder et al. [13] have described a case of H_2S poisoning in a remote rural area of the United States. First responders recognized the "rotten egg" smell and implemented a Hazmat procedure to extricate the man; he was then taken by ambulance to the nearest hos-pital (a 22-minute ride) and then transferred to a burn facility by helicopter (a 38-minute ride). Ambulance personnel wore personal protective equip-ment (PPE) and reported no adverse effects. Helicopter personnel, who did not wear PPE, complained of watery eyes, headache, and dizziness. The hos-pital personnel, who did not use PPE, experienced watery eyes and headache.

During the H_2S suicide fad in Japan several years ago, paramedics and health workers were sometimes injured in a similar manner (see Section 5.2.3.1). And the same has been noted in other countries to which the suicide craze spread. In February 2010 in St. Petersburg, Florida, first responders answered a call about a man unconscious in a car. They noted warning signs posted on the car and donned full protective clothing and breathing devices. Despite these precautions, one policeman inhaled a quantity of H_2S and had to be hospitalized.

3.2 Why H_2S Is So Lethal

Hydrogen sulfide is extremely toxic. But what is it about hydrogen sulfide—in addition to its toxicity—that makes it such a hazard? There are some char-acteristics that make H_2S extremely dangerous:

- The exposure–response curve is extremely steep.
- It can induce instant loss of consciousness at fairly low levels (from 250 ppm), eliminating the victim's chance to flee.
- It tends to accumulate in confined or enclosed spaces.
- It goes suddenly from dissolved to gaseous (soda-can effect), so sudden clouds of H_2S can arise above sewage or stagnant water.
- For all practical purposes, it has no warning odor. However, there is a common misperception that it does.

3.2.1 Exposure–Response Curve

The exposure–response curve for H_2S lethality, shown in Figure 3.1, is extremely steep. At 1000 ppm, death occurs in slightly less than 10 seconds. At 300 ppm, death occurs after a few hours.

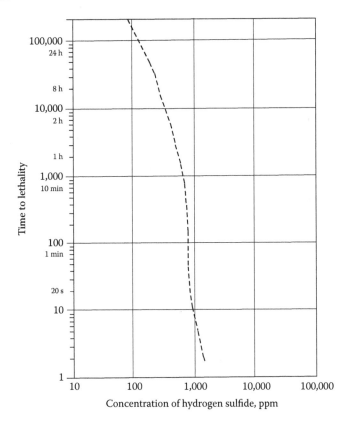

FIGURE 3.1

Exposure–response curve for H_2S lethality. (Reprinted from Guidotti, T.L., *Occup. Med.*, 46(5), 367, 1996. With permission. Adapted by Guidotti from an unpublished report by Dr. Robert Rogers for the Alberta Energy Resources Conservation Board, 1990.)

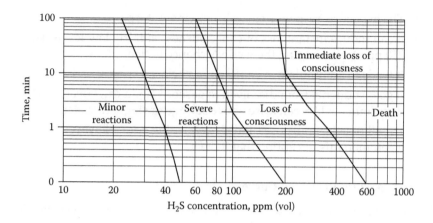

FIGURE 3.2

Physiological effects of H₂S. (Reprinted from Gillies, A.D.S. et al., The modelling of the occurrence of hydrogen sulfide in coal seams, *Proceedings of the Eighth US Mine Ventilation Symposium*, Rolla, MO, Society for Mining, Metallurgy and Exploration, Englewood, CO, 1999, pp. 709–720. With permission.)

Unfortunately, 300 ppm can prove to be fatal, as the loss of consciousness, which can occur at 250 ppm and upward, will reduce the chances of flight or removal from the danger zone [8].

Exposure guidelines for poisonous gases are often set by applying Haber's law. Haber's law states that the severity of the toxic effect depends on the total exposure: $c \times t$, where c is the gas concentration and t is the exposure time [14,15].

H₂S does not follow Haber's law. Instead, the concentration is the primary determinant: higher concentrations are much more toxic, even at proportionally shorter exposure times. A short exposure at high concentrations simply will not equate to a long exposure at lower concentration. For H₂S, toxicity is driven by concentration.

Many current models for H₂S lethality use a nonlinear equation, such as $C^n \times t$, where C is concentration, t is exposure time, and n ranges from 1.4 to 4.36 or higher [5].

The concentration-driven toxicity of H₂S is elegantly shown by Gillies et al. [16] in Figure 3.2. In this diagram, it can be seen that exposure time has a certain effect, but this is minor compared to the concentration.

3.2.2 H₂S in Enclosed Spaces

Hydrogen sulfide is slightly heavier than air, so it accumulates in subterranean or low-lying confined spaces, excavations, and depressions. Valve chambers and sewers are classic sites where H₂S accumulates.

In April 1969, at Portage La Prairie in Canada, a three-member crew went to the sewage lagoons to check a feeder line valve. One member of the crew entered the valve chamber and collapsed. The second member went in to help the first and also collapsed. The third person summoned help. Both workers in the valve chamber were dead when the fire department arrived. Air samples taken at the time showed high levels of H_2S in the valve chamber [7].

3.2.3 Soda-Can Effect

H_2S is highly water soluble—approximately 10 times more so than carbon dioxide (see Table 3.1)—so unstirred water can contain large quantities. When the water is disturbed, the dissolved H_2S can quickly enter the gaseous phase and form a lethal cloud—the "soda can" effect.

As the ambient temperature increases, the solubility decreases slightly; however, the speed of the "soda can" effect increases. Yongsiri et al. [17] have shown that the transfer rate of H_2S from the aqueous phase to the gas phase increases with increasing temperature.

3.2.4 No Warning Odor

This gas has a very distinctive smell of rotten eggs. Many people have smelled the characteristic odor at low concentrations and not suffered ill effects. Ironically, the rotten egg smell may actually encourage complacency: a common misapprehension is that "as long as you can smell it, you're not in danger."

The problem is that humans can only detect the odor of H_2S at very low concentrations and only for a limited time. With continuous low-level exposure, the ability to smell H_2S is often lost. At high concentrations, the loss of ability to smell H_2S happens instantly. So just as levels become dangerous, the ability to detect H_2S by sense of smell disappears.

TABLE 3.1

Solubility of H_2S and CO_2 in Water

Temperature (°C)	Solubility, H_2S	Solubility, CO_2
15	2.335×10^{-3}	8.21×10^{-4}
20	2.075×10^{-3}	7.07×10^{-4}
25	1.85×10^{-3}	6.15×10^{-4}
30	1.66×10^{-3}	5.41×10^{-4}
35	1.51×10^{-3}	4.80×10^{-4}

Source: Modified from *Handbook of Chemistry & Physics*, 76th edn., 1995–1996, CRC Press, pp. 6–4.

Detection of H_2S through odor is *not* reliable. For all practical purposes, it should be treated as a hazard that gives no warning odor [18].

3.3 Exposure Effects and Permissible Limits

Table 3.2 shows the dose–response data found in the literature for H_2S intoxication.

3.3.1 Interpreting the Dose–Response Data

Some trends can be seen in the data, even if the numbers must be viewed as approximate:

- At 0–50 ppm, irritant effects predominate. The *duration* of each exposure and the number of exposures over time (chronicity) are important in determining the severity and permanence of the symptoms.
- At 200–500 ppm, systemic H_2S intoxication occurs. The duration of exposure is critical at this concentration. Rapid intervention can save life, but a few hours at this concentration will extinguish life.
- At 700 ppm or higher, H_2S intoxication tends to be immediate and fatal. Death results from respiratory center paralysis, asphyxia, and cardiac failure.

The concentration boundaries must be viewed as quite rough; irritant effects, for example, might be described as dominating over the range 0–100 ppm instead of our conservative 0–50 ppm. And the ranges 100–200 ppm and 500–700 ppm are quite frankly gray areas, possibly because these are transition zones from one mechanism to another.

The overall picture, on a logarithmic scale, might be something like the diagram in Figure 3.3.

3.3.1.1 Olfactory Fatigue versus Olfactory Paralysis

In Table 3.2 the terms "olfactory fatigue" and "olfactory paralysis" may appear to be used interchangeably. However, as Guidotti [5] points out, these are two distinct phenomena. Olfactory fatigue is a general—though usually incomplete—adjustment to an intense odor. Olfactory paralysis is a direct neurotoxic effect. Olfactory paralysis is more serious, because it nullifies the most important warning sign that exposed people might have.

TABLE 3.2

H$_2$S Dose–Effect Relationships Described in the Literature

Concentration (Parts per Million)	Effect	References
0.007	Odor threshold	[19]
Circa 0.03	Odor threshold	[20]
0.02–0.13	Odor threshold	[21]
0.01–0.3	Odor threshold	[22]
0.13	Unpleasant odor is noticeable. Sore eyes	[7]
0.15	Offensive odor	[23]
0.2	Detectable odor	[24]
0.77	Faint odor	[25]
1.88	Bronchial constriction in asthmatic individuals	[26]
2.5–5	Coughing, throat irritation	[27]
3.4	Increased eye complaints	[28]
4.6	Strong intense odor, but tolerable. Prolonged exposure may deaden the sense of smell	[7]
3.4–19	Eye irritation	[29]
1–20	Offensive odor, possible nausea, tearing of the eyes or headaches with prolonged exposure	[22]
5–10	Eye irritation is possible	[30]
10	Offensive odor	[31]
10–20	Threshold for eye irritation	[32]
10–20	Causes painful eye, nose, and throat irritation, headaches, fatigue, irritability, insomnia, gastrointestinal disturbance, loss of appetite, dizziness. Prolonged exposure may cause bronchitis and pneumonia	[7]
18.8	Fatigue, loss of appetite, headache, irritability, poor memory, dizziness	[23]
20–50	Nose, throat, and lung irritation; digestive upset and loss of appetite; sense of smell starts to become fatigued; acute conjunctivitis may occur (pain, tearing, and light sensitivity)	[22]
30	Inflammation of the eyes	[33]
30–100	Sickeningly sweet smell noted	[7]
20–30	Intense odor	[21]
50	Ability to smell H$_2$S can begin to dull	[34]
50	May cause muscle fatigue, inflammation, and dryness of nose, throat, and tubes leading to the lungs. Exposure for 1 h or more at levels above 50 ppm can cause severe eye tissue damage. Long-term exposure can cause lung disease	[7]
50	Conjunctivitis with ocular pain, lacrimation, and photophobia. Can progress to keratoconjunctivitis and vesiculation of the corneal epithelium	[35]

(Continued)

TABLE 3.2 (*Continued*)

H$_2$S Dose–Effect Relationships Described in the Literature

Concentration (Parts per Million)	Effect	References
50	Conjunctival irritation is noticeable	[20,36]
50	Conjunctivitis ("gas eyes"), and upper respiratory irritation over time	[23]
50–100	Serious eye damage	[32]
50–100	Irritation of eyes and respiratory tract	[37]
50–100	Mild mucous membrane irritation; nausea, vomiting, corneal ulceration, keratoconjunctivitis	[38]
>94	Olfactory paralysis	[39]
100	Olfactory fatigue	[40,41]
100	Olfactory fatigue; eyes and throat may sting	[31]
100–150	Olfactory fatigue/paralysis	[38]
150	Olfactory nerve paralysis	[20,23,24]
100–150	Loss of smell, stinging of eyes and throat. Fatal after 8–48 hours of continuous exposure	[7]
100–200	Severe nose, throat, and lung irritation; ability to smell odor completely disappears	[22]
200	Intense stinging of eyes and throat; smell disappears rapidly	[31]
150–250	Loss of olfactory sense	[32]
>200	Pulmonary edema on prolonged inhalation	[42]
200–250	Nervous system depression (headache, dizziness, and nausea are symptoms). Prolonged exposure may cause fluid accumulation in the lungs. Fatal in 4–8 hours of continuous exposure	[7]
200–300	Rhinitis, bronchitis, pulmonary edema	[38]
240	Unconsciousness after a period (not recorded how long)	[43]
250	Pulmonary edema	[44]
250–500	Pulmonary edema	[22]
250–600	Pulmonary edemas (lungs fill with fluid, foaming in mouth, chemical damage to lungs)	[7]
300	May cause muscle cramps, low blood pressure, and unconsciousness after 20 minutes. 300 to 500 ppm may be fatal in 1–4 hours of continuous exposure	[7]
300–500	Pulmonary edema, imminent threat to life	[45]
320–530	Pulmonary edema with risk of death	[29]
>375	Respiratory distress	[46]
>469	Death	[21]
500	Severe lung irritation, excitement, headache, dizziness, staggering, sudden collapse (knockdown), unconsciousness and death within a few hours, loss of memory for the period of the exposure	[22]

(Continued)

TABLE 3.2 (*Continued*)

H$_2$S Dose–Effect Relationships Described in the Literature

Concentration (Parts per Million)	Effect	References
>500	Lesional pulmonary edema; coma, possibly with convulsions	[41]
500	Causes excitement, headache, dizziness, and staggering, followed by unconsciousness and respiratory failure	[37]
500	Systemic symptoms can appear	[24]
500	Paralyzes the respiratory system and overcomes victim almost instantaneously. Death after exposure of 30–60 minutes	[7]
500	Dizziness. Breathing stops in a few minutes; prompt artificial respiration needed	[31]
500	Unconsciousness in a few seconds	[9]
500–700	Clinical signs consistent with acute toxicosis	[45,47]
500–700	Unconsciousness within seconds; cardiopulmonary arrest secondary to asphyxia and respiratory paralysis	[38]
500–1000	Acts primarily as a systemic poison causing unconsciousness and respiratory failure	[48]
500–1000	Respiratory paralysis, irregular heartbeat, collapse, and death without rescue	[22]
530–1000	Strong central nervous system stimulation, hyperpnoea followed by respiratory arrest	[29]
650–700	Rapid loss of consciousness followed by permanent severe neurological sequelae (chronic vegetative state)	[49]
700	Unconscious quickly, death results if not rescued promptly	[31]
700	Paralysis of the nervous system	[7]
700	Cardiopulmonary arrest	[50]
700–900	Rapid loss of consciousness, apnea (central respiratory paralysis)	[25]
750–1000	Abrupt physical collapse; depending on how pronounced and/or prolonged exposure is, collapse can give way to rapidly fatal respiratory paralysis	[42,51,52]
>900	Instant death can result	[37]
1000	Rapid collapse, respiratory paralysis imminent	[24]
1000	Nervous system paralysis	[23]
>1000	Near-instant respiratory paralysis and coma	[12]
>1000	Rapid collapse and death	[22]
1000	Unconscious at once, dead within minutes	[31]
1000	Immediately fatal	[7]
1000–2000	Immediate collapse with paralysis of respiration	[29]
5000	Imminent death	[24]

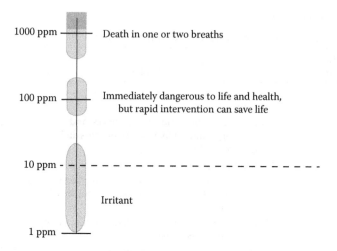

1000 ppm ——— Death in one or two breaths

100 ppm ——— Immediately dangerous to life and health,
                  but rapid intervention can save life

10 ppm - —— -

              Irritant

1 ppm ———

FIGURE 3.3

Hydrogen sulfide exposure as logarithmic scale. The broken line at 10 ppm indicates the threshold limit value–time-weighted average used in many jurisdictions.

3.3.2 Exposure Limits

Various authorities have differing permissible levels for airborne hydrogen sulfide, as shown in Table 3.3.

The ACGIH's threshold limit value (TLV®) for H_2S is 1 ppm, 8-hour time-weighted average (before 2010, 10 ppm), and 5 ppm short-term exposure limit (before 2010, 15 ppm).

The National Institute for Occupational Safety and Health (NIOSH) has determined the concentration that should be regarded as immediately dangerous to life or health to be 100 ppm. NIOSH has also set a Recommended Exposure Limit (REL) of 10 ppm, which should not be exceeded during any 10-minute work period.

3.3.2.1 Community versus Workplace Standards

Community standards or odor nuisance statutes are intended to prevent or ameliorate smells that interfere with the well-being of community residents. Odors that cause people to stay indoors, or temporarily leave the neighborhood, or are annoying or irritating can qualify as a nuisance, even if it is below the levels that can cause medical symptoms [53,54]. Many countries do not have ambient air quality levels for H_2S, since it is not perceived as a community problem in most regions.

It may seem, at first glance, that the workplace levels are extraordinarily high compared to community standards. They are based upon very different considerations: the workforce is assumed to be adults in generally good health and working 8-hour per day (40 hours per week). People living in the community represent all ages and all states of health; and they spend most, or even all, of the 24 hours of the day, in the community.

TABLE 3.3

Permissible Levels for H$_2$S

Jurisdiction	Exposure Type	Description
United States	Workplace, general industry	Exposures shall not exceed 20 ppm (ceiling) with the following exception: if no other measurable exposure occurs during the 8-hour work shift, exposures may exceed 20 ppm, but not more than 50 ppm (peak), for a single time period up to 10 minutes
United States	Workplace, construction industry	Time-weighted average (TWA), 10 ppm
EU (SCOEL[a])	Workplace	TWA (8 hours), 5 ppm; STEL (15 minutes), 10 ppm
Australia	Workplace	TWA, 10 ppm; STEL, 15 ppm
Canada—through 2009	Workplace	TWA, 10 ppm; STEL, 15 ppm
Canada—from 2010	Workplace	Limits set per province. Some provinces automatically follow the American Conference of Governmental Industrial Hygienists (ACGIH) TLV and have therefore 1 ppm TWA and 5 ppm STEL. Others, e.g., Labrador and Newfoundland, follow the 2009 ACGIH TLVs
WHO	Community	0.11 ppm averaged over 24 hours (guideline)
State of California	Community	0.03 ppm averaged over 1 hour
State of Hawaii	Community	0.025 ppm averaged over 1 hour
New Zealand	Community	0.005 ppm averaged over 1 hour (guideline)

[a] SCOEL is the European Union's equivalent to NIOSH.

3.4 Chemical and Physical Properties of H$_2$S

Some of the physical properties of H$_2$S contribute to its lethality [55,56]:

- Specific gravity (relative density) = 1.19. Because it is heavier than air, H$_2$S collects in low points such as underground chambers.
- Vapor pressure = 18.75×10^5 Pa. Although it is water soluble, H$_2$S will easily leave the aqueous phase for the gas phase.
- Lipid solubility. The highly lipophilic nature of H$_2$S allows it to easily penetrate the lipid bilayer of cell membranes. It is quickly absorbed through the lungs' alveoli into the bloodstream.

3.4.1 Explosivity/Flammability

In addition to being highly toxic, hydrogen sulfide is both flammable and explosive. The upper and lower explosive limits are shown in Table 3.4.

TABLE 3.4

Explosive Range, H_2S (% in Air)

Parameter	Value (%)
Lower explosive limit (LEL)	4.0
Upper explosive limit (UEL)	46.0

Source: Data from National Institute for Occupational Safety and Health, *NIOSH Pocket Guide to Chemical Hazards*, DHHS Publication No. 2005-149, U.S. Government Printing Office, Pittsburgh, PA, 2007.

Hydrogen sulfide is a strong reducing agent and highly reactive. In atmospheres where the concentration of H_2S is greater than the oxygen concentration, the H_2S can react with rust or other corrosion products to produce iron sulfide. Iron sulfide is pyrophoric—it can ignite spontaneously when exposed to air [22]. Oxidants such as fluorine, sodium peroxide, or copper chromate ignite upon contact. Some chemicals such as dichlorine oxide, copper powder, or soda lime will explode upon contact with H_2S [57].

3.4.2 Other Names for H_2S

Other names for H_2S are dihydrogen sulfide, dihydrogen monosulfide, sulfur hydride, hydrogen sulfuric acid, hydrosulfuric acid, sulfurated hydrogen, and hepatic gas; also, less scientifically, "stink damp," "sour gas," and "sewer gas" [7,22,58–61].

3.4.3 Conversion Factors

- 1% vol = 10,000 ppm
- 1 mg/L= 717 ppm at STP
- 1 ppm = 1.4 mg/m³ = 1400 μg/m³
- 1 μmol/mL = 32 μg/mL (sulfide in blood)
- 1 μmol/mL = 112 μg/mL (thiosulfate in blood or urine)

3.5 Generation of H_2S

The decomposition of organic matter under anaerobic conditions generates the quantities of H_2S that cause hazards in our industry.

The formation of H_2S is favored by [62]

- High temperature
- Low pH

- Long retention times
- Low velocity in the pipes
- Scarcity of oxygen

3.5.1 Anaerobic Bacteria

The decomposition processes in the sewer are complex; a brief overview is given in Figure 3.4. The breakdown of organic matter takes place in the water, in the sediment, and in the biofilm on the sewer pipe walls. A very simplified explanation might be as follows [63–66].

Wastewater is subjected to alternating aerobic–anaerobic conditions in the sewer network. Inside the pressure mains, conditions tend to be anaerobic. Inside the gravity lines, conditions can be either aerobic or anaerobic, depending on the degree of reaeration during transport.

In general, in aerobic conditions the organic matter is broken down through oxidation by aerobic bacteria. This consumes the dissolved oxygen (DO). When the DO is used up, the conditions turn to anaerobic; the bacterial population in the biofilm adapts accordingly.

The new florae of bacteria contain more sulfur-reducing bacteria. The sulfur-reducing bacteria transform the sulfates (SO_4^{2-}) in the water to sulfides (S^{2-}) and then to hydrogen sulfide.

H_2S production increases with increasing temperatures, high chemical oxygen demand (COD) concentrations, longer retention times, and pipe diameter.

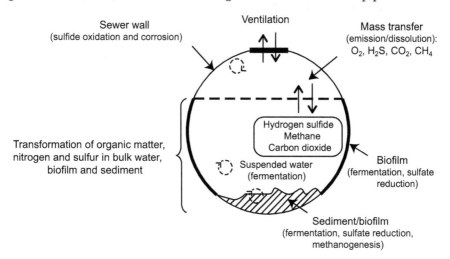

In-sewer processes

FIGURE 3.4
In-sewer processes that generate H_2S. (Reprinted from Lim, J.S. et al., *Ad Hoc Netw.*, 11(4), 1456, 2013. With permission.)

3.5.2 Endogenous H₂S

The only aspect of H_2S that is relevant for the wastewater industry is its toxicity and the dangers it presents to industry workers.

Scientific advances have been made recently in the understanding of endogenous H_2S, and this subject has received a good deal of attention in the technical literature. To avoid possible confusion, we would like to briefly mention what endogenous H_2S is and why it has nothing to do with our industry.

Endogenous H_2S are molecules that the human body produces and uses for certain processes. The amounts of H_2S involved in these biological processes are orders of magnitude lower than the amounts with which we are concerned. A consensus seems to have formed that molecules of H_2S are generated by vascular smooth muscle cells and used as a gaseous signaling compound (gasotransmitter), along with the previously identified gasotransmitters nitrous oxide (NO) and carbon monoxide (CO) [67–70]. Endogenous H_2S is also believed to play a role in inflammatory responses [71–73].

We would like to reiterate that the amounts of endogenous H_2S generated by the body are miniscule compared to the amounts with which the wastewater industry is concerned.

3.6 Scope of the Problem

3.6.1 Affected Groups

It has been known for a long time that H_2S can occur more or less continuously in some industries:

- Sewage transport and treatment
- Petroleum and natural gas production and refining
- Liquid manual pits in farming
- Pulp mills (kraft process)
- Viscose rayon production
- Rubber vulcanization
- Heavy water production
- Industrial fishing, fish processing, and fish farming
- Geothermal energy installations

The first two categories are the highest contributors to H_2S poisonings. Hendrickson et al. [74] examined the U.S. Bureau of Labor Statistics' Census of Fatal Occupational Injuries for deaths related to H_2S for the 1993–1999 period.

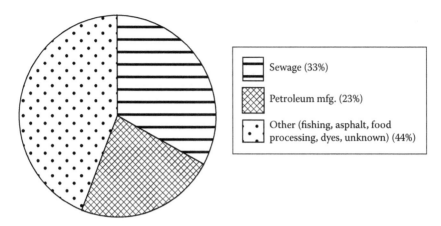

Sewage (33%)

Petroleum mfg. (23%)

Other (fishing, asphalt, food processing, dyes, unknown) (44%)

FIGURE 3.5
U.S. occupational H_2S poisonings 1993–1999, by industry. (Data from Hendrickson, R.G. et al., *Am. J. Ind. Med.*, 45(4), 346, 2004.)

They found that the first two industries in the list above accounted for more than half—56%—of all H_2S poisonings (see Figure 3.5).

There can be overlap between these industries, since H_2S is produced by decaying organic matter, and organic matter is sometimes disposed of via sewer systems. Wastewater treatment plants (WWTP) that treat industrial wastewater can be exposed to sudden spikes of H_2S; Tvedt has described an H_2S poisoning at a WWTP, where decaying offal from an upstream shrimp processing plant caused an unexpected spike of H_2S at the downstream WWTP [75].

Nikkanen and Burns [76] have reported on fatal H_2S accidents from cleaning tanks at a fish hatchery. In that case, the H_2S poisoning occurred at the fish hatchery. However, it is not inconceivable that had the tanks been cleaned by flushing with water, then the problem would have moved downstream to the sewer system or the WWTP.

There are also natural sources of H_2S, mainly certain mineral springs, volcanic gases, and rock fissure gases; and it can arise from bacterial action in brackish waters [36].

3.6.2 How Many People?

In the province of Alberta (Canada), with its heavy concentration of high-sulfur-content oil and gas fields, an average of two deaths per year have been reported from H_2S exposure in the oil and gas industry [52,77].

Fuller and Suruda [78] reviewed the U.S. Occupational Safety and Health Administration (OSHA) investigation records for the period 1984–1994 and found 80 fatalities from H_2S in 57 incidents; 19 fatalities were coworkers attempting to rescue fallen workers. In 60% of the fatalities, OSHA issued citations for violation of respiratory protection and confined space standards. The authors note that the use of H_2S detection

TABLE 3.5

Average Number of H_2S Fatalities per Year, from Various Studies

Location	Years	Average Number of Fatalities/Year	Notes	References
United States	1999–2007	5		[80]
United States	2007	13		[81]
United States	2005	6	1396 cases of H_2S exposure reported.	[82]
United States	1993–1999	7.4	Up to 21% of fatalities were coworkers attempting to rescue.	[74]
United States	1992–1998	7.6	Sulfur compounds— H_2S, SO_2, etc.	[79]
United States	1984–1994	7.3	24% of fatalities were coworkers attempting to rescue.	[78]
Alberta province, Canada	1979–1983	1.4	250 cases of H_2S exposure in this period. Overall mortality = 2.8%.	[77]
Alberta province, Canada	1969–1973	2.6	221 cases of H_2S exposure in this period (75% petrochemical industry, 25% sewers or pumping stations). Overall mortality = 6%.	[50]

equipment, air-supplied respirators, and confined space safety training would have prevented most of the fatalities.

In their review of the U.S. Bureau of Labor Statistics' Census of Fatal Occupational Injuries (CFOI) for deaths related to H_2S for the 1993–1999 period, Hendrickson et al. identified 52 deaths due to H_2S poisoning. Working independently from the CFOI database for 1992–1998, Valent et al. reached similar conclusions (see Table 3.5). In 21% of the cases, a coworker died simultaneously or in a rescue attempt [74,79].

Arnold et al. [77] note that in Alberta province, the overall mortality in H_2S exposures dropped from 6% in the 1969–1973 period to 2.8% in the 1979–1983 period. They attribute this to improved first-aid training and increased awareness of the dangers of H_2S.

3.6.3 Company Size, Length of Employment Are No Guarantees

It is sometimes seen in the literature that those workers most liable to be victims of H_2S intoxication are new workers at small companies. This may seem reasonable; however, at least one study has found that company size and length of employment are no guarantee when it comes to H_2S.

When examining the U.S. Bureau of Labor Statistics' CFOI for 1993–1999, Hendrickson et al. [74] found that small-, medium-, and large-sized companies are all vulnerable (see Figure 3.6).

In the same study, Hendrickson et al. report that workers employed less than 1 year made up nearly half the fatalities. This seems reasonable: new workers may simply fail to recognize the risk in many situations. "Learning by doing" is *not* to be recommended when it comes to H_2S safety. However, as Figure 3.7 shows, experience is no guarantee either; slightly more than half of those killed had been employed more than 1 year.

We believe that Figures 3.6 and 3.7 prove the need for both early training for new employees and regular follow-up or "brush-up" training for experienced employees.

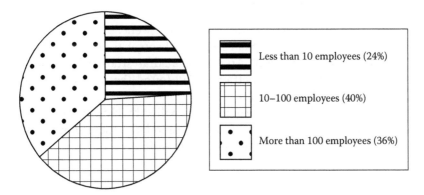

FIGURE 3.6
U.S. occupational H_2S fatalities 1993–1999, by company size. (Data from Hendrickson, R.G. et al., *Am. J. Ind. Med.*, 45(4), 346, 2004.)

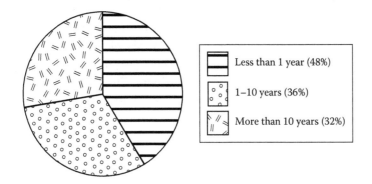

FIGURE 3.7
U.S. occupational H_2S fatalities 1993–1999, by years of employment. (Data from Hendrickson, R.G. et al., *Am. J. Ind. Med.*, 45(4), 346, 2004.)

3.7 The Myth of Harmless Knockdown

It is sometimes stated in the literature, especially older reports, that if the victim of H_2S poisoning survives, then recovery is prompt and complete [42,51]. There are very many documented cases proving this to be a false assumption [6,49,83]. It is possible that this confusion is the result of conflating two things:

- A very large fraction of fatalities occur on the scene.
- "Knockdown"—immediate, temporary unconsciousness caused by H_2S.

3.7.1 Fatalities Occurring on the Scene

A very large fraction of fatalities do occur on the scene. In their examination of labor fatalities in the United States, Hendrickson et al. [74] report that 87% of H_2S deaths occur on the spot. And in a Canadian study of H_2S injuries in Alberta during 1969–1973, Burnett et al. [50] found that 71% of H_2S deaths occurred before reaching hospital.

To be brutally frank: if you are going to be killed by H_2S, the odds are that it will happen quickly, before you reach a hospital. But that is not to say that if you aren't killed, then you will suffer no aftereffects.

3.7.2 "Knockdown"

In the oil and natural gas industry, there have been reported several cases of workers losing consciousness due to H_2S and waking up with apparently no ill effects. We should like to make some observations about this:

- Knockdowns have been known to be fatal if exposure is prolonged, probably as a result of respiratory paralysis [8].
- Due to the outdoor nature of work in the oil and gas industry, many knockdown incidents occur out of doors. The H_2S might be dissipated by air movement before significant injury occurs.
- It has been known to happen that workers fall out of the H_2S cloud when they lose consciousness. This effectively removes them immediately from the source of H_2S, before significant hypoxic injury occurs. For a description of such an event, see Gabbay et al. [38].
- Epidemiology studies show that those who have experienced knockdown are more liable to suffer effects [5,84].

References

1. McAnalley, B. H., Lowry, W. T., Oliver, R. D., and Garriott, J. C. (1979). Determination of inorganic sulfide and cyanide in blood using specific ion electrodes: Application to the investigation of hydrogen sulfide and cyanide poisoning. *Journal of Analytical Toxicology*, 3(3), 111–114.
2. Spiers, M. and Finnegan, O. C. (1986). Near death due to inhalation of slurry tank gases. *The Ulster Medical Journal*, 55(2), 181.
3. Greenberg, M. and Hamilton, R. (1998). The epidemiology of deaths related to toxic exposures in the US workplace, 1992–1996. *Journal of Toxicology: Clinical Toxicology*, 5, 430.
4. Policastro, M. A. and Otten, E. J. (2007). Case files of the University of Cincinnati fellowship in medical toxicology: Two patients with acute lethal occupational exposure to hydrogen sulfide. *Journal of Medical Toxicology*, 3(2), 73–81.
5. Guidotti, T. L. (2010). Hydrogen sulfide advances in understanding human toxicity. *International Journal of Toxicology*, 29(6), 569–581.
6. Kemper, F. D. (1966). A near-fatal case of hydrogen sulfide poisoning. *Canadian Medical Association Journal*, 94(21), 1130.
7. CUPE. (1987). Health and safety fact sheet: Hydrogen sulfide. CUPE Publication No. 491. CUPE National Health and Safety Branch, Canadian Union of Public Employees (CUPE): Ottawa, Ontario, Canada.
8. Guidotti, T. L. (1996). Hydrogen sulphide. *Occupational Medicine*, 46(5), 367–371.
9. Chou, C.-H. S. J. (2003). Hydrogen sulfide: Human health aspects. Concise international chemical assessment document 53. World Health Organization: Geneva, Switzerland.
10. Sewer Entry Guidelines. (2010). *Workplace Health and Safety Bulletin CH037*. Queen's Printer, Government of Alberta, Department of Employment and Immigration: Edmonton, Alberta, Canada.
11. Gray, C., Vaughn, J. L., and Sanger, K. (2011). Distribution/collection certification study guide. Oklahoma State Department of Environmental Quality: Oklahoma City, OK.
12. Knight, L. D. and Presnell, S. E. (2005). Death by sewer gas: Case report of a double fatality and review of the literature. *The American Journal of Forensic Medicine and Pathology*, 26(2), 181–185.
13. Ruder, J. B., Ward, J. G., Taylor, S., Giles, K., Higgins, T., and Haan, J. M. (2015). Hydrogen sulfide suicide: A new trend and threat to healthcare providers. *Journal of Burn Care and Research*, 36(2), e23–e25.
14. Gaylor, D. W. (2000). The use of Haber's law in standard setting and risk assessment. *Toxicology*, 149(1), 17–19.
15. Hoyle, G. W., Chang, W., Chen, J., Schlueter, C. F., and Rando, R. J. (2010). Deviations from Haber's Law for multiple measures of acute lung injury in chlorine-exposed mice. *Toxicological Sciences*, 118(2): 696–703.
16. Gillies, A. D. S. et al. (June 1999). The modelling of the occurrence of hydrogen sulphide in coal seams. In: *Proceedings Eighth US Mine Ventilation Symposium*, Rolla, MO. Society for Mining, Metallurgy and Exploration, Englewood, MO, pp. 709–720.

17. Yongsiri, C., Vollertsen, J., and Hvitved-Jacobsen, T. (2004). Effect of temperature on air-water transfer of hydrogen sulfide. *Journal of Environmental Engineering,* 130(1), 104–109.
18. Occupational Safety and Health Administration (OSHA). (2005). Hydrogen sulfide, OSHA Fact Sheet series. U.S. Department of Labor: Washington, DC.
19. Amoore, J. E. and Hautala, E. (1983). Odor as an aid to chemical safety: Odor thresholds compared with threshold limit values and volatilities for 214 industrial chemicals in air and water dilution. *Journal of Applied Toxicology,* 3, 272–290.
20. Evans, C. L. (1967). The toxicology of hydrogen sulfide and other sulfides. *Quarterly Journal of Experimental Physiology and Cognate Medical Sciences,* 52(3), 231.
21. Beauchamp, R. O., Bus, J. S., Popp, J. A., Boreiko, C. J., Andjelkovich, D. A., and Leber, P. (1984). A critical review of the literature on hydrogen sulfide toxicity. *CRC Critical Reviews in Toxicology,* 13(1), 25–97.
22. Hydrogen Sulphide at the Work Site. (2010). *Workplace Health and Safety Bulletin CH029.* Queen's Printer, Government of Alberta, Department of Employment and Immigration: Edmonton, Alberta, Canada.
23. Ahlborg, G. (1951). Hydrogen sulfide poisoning in shale oil industry. *AMA Archives of Industrial Hygiene and Occupational Medicine,* 3(3), 247–266.
24. Gafafer, W. M., ed. (1964). Hydrogen sulfide, in *Occupation Diseases: A Guide to Their Recognition,* Public Health Service Publication No. 1097. U.S. Department of Health, Education and Welfare: Washington, DC, p. 163.
25. Poda, G. A. (1966). Hydrogen sulfide can be handled safely. *Archives of Environment Health,* 12, 795.
26. Jäppinen, P., Vilkka, V., Marttila, O., and Haahtela, T. (1990). Exposure to hydrogen sulphide and respiratory function. *British Journal of Industrial Medicine,* 47(12), 824–828.
27. Bhambhani, Y. and Singh, M. (1991). Physiological effects of hydrogen sulfide inhalation during exercise in healthy men. *Journal of Applied Physiology,* 71(5), 1872–1877.
28. Vanhoorne, M., De Rouck, A., and De Bacquer, D. (1995). Epidemiological study of eye irritation by hydrogen sulfide and/or carbon disulphide exposure in viscose rayon workers. *Annals of Occupational Hygiene,* 3, 307–315.
29. WHO. (1981). Environmental Health Criteria, No. 19: Hydrogen Sulfide. World Health Organization (WHO): Geneva, Switzerland. International Programme on Chemical Safety.
30. Elkins, H. B. (1950). *The Chemistry of Industrial Toxicology.* John Wiley & Sons: New York, pp. 95 and 232.
31. Jones, J. P. (1975). Hazards of hydrogen sulfide gas. *Sel. Paper 23rd Annual Gas Meas. Inst.,* 16.
32. Savolainen, H. (1982). Nordiska expertgruppen för gränsvärdesdokumentation. 40. Dihydrogensulfid. *Arbete och Hälsa,* 31, 1–27.
33. Barthelemy, H. L. (1939). Ten years' experience with industrial hygiene in connection with the manufacture of viscose rayon. *Journal of Industrial Hygiene and Toxicology,* 21, 141–151.
34. Pouliquen, F. et al. (2005). Hydrogen sulfide. In: *Ullmann's Encyclopedia of Industrial Chemistry,* 7th ed. John Wiley & Sons: New York.
35. ACGIH. (1991). *Documentation of the Threshold Limit Values and Biological Exposure Indices,* 6th ed. American Conference of Governmental Industrial Hygienists: Cincinnati, OH, Vol. II, pp. 786–788.

36. Yant, W. P. (June 1930). Hydrogen sulphide in industry—Occurrence, effects, and treatment. *American Journal of Public Health and the Nations Health*, 20(6), 598–608.
37. Johnstone, R. T. and Miller, S. E. (1960). Noxious gases: Hydrogen sulfide (H_2S), in *Occupational Diseases and Industrial Medicine*. W.B. Saunders: Philadelphia, PA, p. 115.
38. Gabbay, D. S., De Roos, F., and Perrone, J. (2001). Twenty-foot fall averts fatality from massive hydrogen sulfide exposure. *The Journal of Emergency Medicine*, 20(2), 141–144.
39. Hirsch, A. R. and Zavala, G. (1999). Long term effects on the olfactory system of exposure to hydrogen sulfide. *Occupational and Environmental Medicine*, 56, 284–287.
40. NIOSH. (May 1977). Criteria for a recommended standard, occupation exposure to hydrogen sulfide. Publication No. 77–158. The National Institute for Occupational Safety and Health, U.S. Department of Health, Education and Welfare: Washington, DC.
41. Christia-Lotter, A., Bartoli, C., Piercecchi-Marti, M. D., Demory, D., Pelissier-Alicot, A. L., Sanvoisin, A., and Leonetti, G. (2007). Fatal occupational inhalation of hydrogen sulfide. *Forensic Science International*, 169(2), 206–209.
42. Milby, T. H. (1962). Hydrogen sulfide intoxication: Review of the literature and report of unusual accident resulting in two cases of nonfatal poisoning. *Journal of Occupational and Environmental Medicine*, 4(8), 431–437.
43. Napoli, A. M., Mason-Plunkett, J., Valente, J., and Sucov, A. (2006). Full recovery of two simultaneous cases of hydrogen sulfide toxicity. *Hospital Physician*, 42(5), 47.
44. Tanaka, S., Fujimoto, S., Tamagaki, Y., Wakayama, K., Shimada, K., and Yoshikawa, J. (1999). Bronchial injury and pulmonary edema caused by hydrogen sulfide poisoning. *The American Journal of Emergency Medicine*, 17(4), 427–429.
45. Haggard, H. W. (1925). The toxicology of hydrogen sulfide. *Journal of Indian Hygiene*, 7, 113.
46. Spolyar, L. W. (1951). Three men overcome by hydrogen sulfide in starch plant. *Industrial Health Monthly*, 11, 116–117.
47. Witte, S. T. (1993). Diagnosing hydrogen sulfide toxicosis with a silver/sulfide ion-selective electrode. Doctoral dissertation, Department of Veterinary Pathology, Iowa State University: Ames, IA.
48. ACGIH. (1980). Hydrogen sulfide. In: *Documentation of the Threshold Limit Values*, 4th ed. American Conference of Governmental Industrial Hygienists: Cincinnati, OH, p. 225.
49. Hoidal, C. R., Hall, A. H., Robinson, M. D., Kulig, K., and Rumack, B. H. (1986). Hydrogen sulfide poisoning from toxic inhalations of roofing asphalt fumes. *Annals of Emergency Medicine*, 15(7), 826–830.
50. Burnett, W. W., King, E. G., Grace, M., and Hall, W. F. (1977). Hydrogen sulfide poisoning: Review of 5 years' experience. *Canadian Medical Association Journal*, 117(11), 1277.
51. Milby, T. H. and Baselt, R. C. (1999). Hydrogen sulfide poisoning: Clarification of some controversial issues. *American Journal of Industrial Medicine*, 35, 192–195.
52. Guidotti, T. L. (1994). Occupational exposure to hydrogen sulfide in the sour gas industry: Some unresolved issues. *International Archives of Occupational and Environmental Health*, 66(3), 153–160.

53. Shusterman, D. (1992). Critical review: The health significance of environmental odor pollution. *Archives of Environmental Health: An International Journal,* 47(1), 76–87.

54. Heaney, C. D., Wing, S., Campbell, R. L., Caldwell, D., Hopkins, B., Richardson, D., and Yeatts, K. (2011). Relation between malodor, ambient hydrogen sulfide, and health in a community bordering a landfill. *Environmental Research,* 111(6), 847–852.

55. Reiffenstein, R. J., Hulbert, W. C., and Roth, S. H. (1992). Toxicology of hydrogen sulfide. *Annual Review of Pharmacology and Toxicology,* 32(1), 109–134.

56. Wang, R. (2012). Physiological implications of hydrogen sulfide: A whiff exploration that blossomed. *Physiological Reviews,* 92(2), 791–896.

57. CHEMINFO Chemical Profile, Hydrogen Sulfide. CHEMINFO Record 313. Canadian Centre for Occupational Health and Safety (CCOHS).: Hamilton, Ontario, Canada. http://www.ccohs.ca/products/databases. Accessed August 7, 2014.

58. Schneider, J. S., Tobe, E. H., Mozley, P. D., Barniskis, L., and Lidsky, T. I. (1998). Persistent cognitive and motor deficits following acute hydrogen sulphide poisoning. *Occupational Medicine,* 48(4), 255–260.

59. Recommendation from the Scientific Committee on Occupational Exposure Limits for Hydrogen Sulphide. (June 2007). SCOEL/SUM/124. Scientific Committee on Occupational Exposure Limits (SCOEL), European Commission: Employment, Social Affairs & Inclusion.

60. Hydrogen sulfide, RTECS # MX1225000. Department of Health and Human Services, Centers for Disease Control and Prevention, National Institute for Occupational Safety and Health. DHHS Registry of Toxic Effects of Chemical Substances (RTECS) database: http://www.cdc.gov/niosh-rtecs/MX12B128.html. Accessed October 18, 2015.

61. Hazardous Substance Fact Sheet. (2012). Hydrogen sulfide. RTK 1017. State of New Jersey Department of Health: Trenton, NJ.

62. Hedmark, P. and Strandberg, T. (2015). Hydrogen sulfide reduction by stripping in a pre-chamber. In: *Conference Proceedings, IWA Odours 2015,* Paris, France, November 2015. The International Water Association: London, UK.

63. Tanaka, N., Hvitved-Jacobsen, T., and Horic, T. (2000). Transformations of carbon and sulfur wastewater components under aerobic-anaerobic transient conditions in sewer systems. *Water Environment Research,* 72, 651–664.

64. Yongsiri, C., Vollertsen, J., and Hvitved-Jacobsen, T. (2005). Influence of wastewater constituents on hydrogen sulfide emission in sewer networks. *Journal of Environmental Engineering,* 131(12), 1676–1683.

65. Bertran de Lis, F., Saracevic, E., and Matsche, N. (2007) Control of sulphide problems in pressure sewers. Sustainable techniques and strategies in urban water management. In: *Novatech 2007 Conference Proceedings.* Groupe de Recherche Rhone-Alpes sur les Infrastructures et l'Eau: Lyon, France, pp. 965–972.

66. Sastre, C., Baillif-Couniou, V., Kintz, P., Cirimele, V., Bartoli, C., Christia-Lotter, M.-A., Piercecchi-Mari, M.-D., Leonetti, G., and Pelissier-Alicot, A.-L. (2013). Fatal accidental hydrogen sulfide poisoning: A domestic case. *Journal of Forensic Sciences,* 58(s1), S280–S284.

67. Abe, K. and Kimura, H. (1996). The possible role of hydrogen sulfide as an endogenous neuromodulator. *The Journal of Neuroscience,* 16(3), 1066–1071.

68. Wang, R. (2002). Two's company, three's a crowd: Can H$_2$S be the third endogenous gaseous transmitter? *The FASEB Journal*, 16(13), 1792–1798.

69. Wang, R. (2003). The gasotransmitter role of hydrogen sulfide. *Antioxidants and Redox Signaling*, 5(4), 493–501.

70. Moore, P. K., Bhatia, M., and Moochhala, S. (2003). Hydrogen sulfide: From the smell of the past to the mediator of the future?. *Trends in Pharmacological Sciences*, 24(12), 609–611.

71. Li, L., Bhatia, M., and Moore, P. K. (2006). Hydrogen sulphide—A novel mediator of inflammation? *Current Opinion in Pharmacology*, 6(2), 125–129.

72. Zhi, L., Ang, A. D., Zhang, H., Moore, P. K., and Bhatia, M. (2007). Hydrogen sulfide induces the synthesis of proinflammatory cytokines in human monocyte cell line U937 via the ERK-NF-κB pathway. *Journal of Leukocyte Biology*, 81(5), 1322–1332.

73. Kang, K., Zhao, M., Jiang, H., Tan, G., Pan, S., and Sun, X. (2009). Role of hydrogen sulfide in hepatic ischemia-reperfusion–induced injury in rats. *Liver Transplantation*, 15(10), 1306–1314.

74. Hendrickson, R. G., Chang, A., and Hamilton, R. J. (2004). Co-worker fatalities from hydrogen sulfide. *American Journal of Industrial Medicine*, 45(4), 346–350.

75. Tvedt, B., Skyberg, K., Aaserud, O., Hobbesland, A., and Mathiesen, T. (1991). Brain damage caused by hydrogen sulfide: A follow-up study of six patients. *American Journal of Industrial Medicine*, 20(1), 91–101.

76. Nikkanen, H. E. and Burns, M. M. (2004). Severe hydrogen sulfide exposure in a working adolescent. *Pediatrics*, 113(4), 927–929.

77. Arnold, I. M., Dufresne, R. M., Alleyne, B. C., and Stuart, P. J. (1985). Health implication of occupational exposures to hydrogen sulfide. *Journal of Occupational and Environmental Medicine*, 27(5), 373–376.

78. Fuller, D. C. and Suruda, A. J. (2000). Occupationally related hydrogen sulfide deaths in the United States from 1984 to 1994. *Journal of Occupational and Environmental Medicine*, 42(9), 939–942.

79. Valent, F., McGwin, G., Bovenzi, M., and Barbone, F. (2002). Fatal work-related inhalation of harmful substances in the United States. *CHEST Journal*, 121(3), 969–975.

80. Reedy, S. J. D., Schwartz, M. D., and Morgan, B. W. (2011). Suicide fads: Frequency and characteristics of hydrogen sulfide suicides in the United States. *Western Journal of Emergency Medicine*, 12(3), 300.

81. Ballerino-Regan, D. and Longmire, A. W. (2010). Hydrogen sulfide exposure as a cause of sudden occupational death. *Archives of Pathology & Laboratory Medicine*, 134(8), 1105.

82. Lai, M. W., Klein-Schwartz, W., Rodgers, G. C., Abrams, J. Y., Haber, D. A., Bronstein, A. C., and Wruk, K. M. (2006). 2005 Annual report of the American Association of Poison Control Centers' national poisoning and exposure database. *Clinical Toxicology*, 44(6–7), 803–932.

83. Lee, E. C., Kwan, J., Leem, J. H., Park, S. G., Kim, H. C., Lee, D. H., Kim, J. H., and Kim, D. H. (2009). Hydrogen sulfide intoxication with dilated cardiomyopathy. *Journal of Occupational Health*, 51(6), 522–525.

84. Hessel, P. A., Herbert, F. A., Melenka, L. S., Yoshida, K., and Nakaza, M. (1997). Lung health in relation to hydrogen sulfide exposure in oil and gas workers in Alberta, Canada. *American Journal of Industrial Medicine*, 31(5), 554–557.

85. Lim, J. S. et al. (2013). SewerSnort: A drifting sensor for in situ wastewater collection system gas monitoring. *Ad Hoc Networks*, 11(4), 1456–1471.
86. (1995–1996). *Handbook of Chemistry & Physics*, 76th ed. CRC Press, pp. 6–4.
87. National Institute for Occupational Safety and Health. (2007). *NIOSH Pocket Guide to Chemical Hazards*, DHHS Publication No. 2005-149. U.S. Government Printing Office, Pittsburgh, PA.

4

Hydrogen Sulfide, Part 2: Toxicology

4.1 Introduction

In Chapter 3, we saw that the overall mortality in H$_2$S poisonings seems to be 5% or less. In this chapter, we will discuss how H$_2$S causes so much damage to the human body and what sequelae the survivors of H$_2$S accidents may face.

As Guidotti [1] says,

> For hydrogen sulfide, as for most toxic agents, characteristic symptoms and signs occur in a pattern, not in isolation or arbitrarily. It is the toxidrome that matters, not the individual symptom or sign.

4.1.1 Broad-Spectrum Toxin

H$_2$S is a broad-spectrum toxin; most organs or systems in the human body are vulnerable to H$_2$S. It is absorbed by the lungs, enters the bloodstream, and spreads rapidly through the body, primarily to the brain, liver, kidney, gut, and pancreas [2,3].

Many body systems may be involved, depending on the concentration and duration of exposure [4]. Because of the number of organs and systems that can be affected, H$_2$S poisoning displays a bewildering range of symptoms.

For survivors of H$_2$S incidents, the most frequent effects are neurological, respiratory, and ophthalmologic [5]. Some myocardial effects have also been reported in the literature; and research has indicted a link between H$_2$S exposure and immune system impairment.

4.1.1.1 Historical Focus on Lethal Doses

Hydrogen sulfide gas is extremely toxic to humans; its toxicity and its ability to render victims unconscious (and therefore unable to retreat to safer atmospheres) can often have fatal outcomes.

Most of our knowledge about H$_2$S exposure–response relationships in humans has been learned from fatal or near-fatal industrial accidents. Traditionally, there has been a focus on lethality in risk assessment models;

as some have pointed out, this can lead to decisions being made on grounds of lethality, rather than sublethal health effects [6–8].

4.1.2 Difficulties in Establishing Dose–Response Relationships

The following are examples of complications in establishing dose–response data:

- Rapid dispersal of H_2S plumes; a lethal concentration can disperse before it is measured.
- Exposure incidents may occur at some significant distance from a toxicological laboratory. In the time between exposure and laboratory analysis, an unknown amount of H_2S metabolism occurs.
- Sulfide ion is unstable and breaks down rapidly in a body, so post-mortem toxicology is difficult [9].
- Synergistic effects from elevated levels of particulates, or other inorganic pollutants, for example, SO_2 and NO_x, on H_2S-induced toxicity [10–13].
- Rescuers are quite properly more concerned with removing workers from the dangerous atmosphere than in quantifying the amounts of H_2S.

4.1.2.1 Linear or Threshold-Type Dose–Response Function?

Another complicating factor is that the dose–response function varies between affected organs or organ systems. Some organs display a near-linear relationship between H_2S concentration and severity of response, while other organs or systems show no response below a certain threshold—but above the threshold, the biological response can be dramatic and overwhelming.

For example, the conjunctival membranes and cornea of the eye show a rather linear response to low levels of H_2S. The responses begin with itching and smarting at the lowest levels; the responses worsen with increasing exposure time and concentration through to permanent vision impairment.

On the other hand, the pulmonary edema caused by H_2S does not seem to be linearly related to the concentration. Instead it shows a threshold-type response at concentrations above 200 ppm [14–16].

4.1.2.2 Examples of Complications in H_2S Accidents

A typical case is described by Christia-Lotter et al. [17]: Two workers are detailed to clear out some sewer pipes; one accidently falls in the sewer manhole before donning his respirator. The colleague alerts the emergency services and a rescue team is on site in 10 minutes. Because of the narrowness of the manhole, it is not possible for rescuers to wear their self-contained

TABLE 4.1

Comparison of Susceptibility of Animals to H_2S

Animal	Approximate Percentage of H_2S for Subacute Symptoms	Approximate Percentage of H_2S for Acute Symptoms
Canaries	0.005–0.020	0.02 or more
White rats	0.005–0.055	0.05 or more
Dogs	0.005–0.065	0.05 or more
Guinea pigs	0.005–0.075	0.075 or more
Goats	0.005–0.090	0.090 or more

Source: Data from Sayers, R.R. et al., Investigation of toxic gases from Mexican and other high-sulfur petroleums and products, *US Bureau of Mines, Bulletin 231*, Government Printing Office, Washington, DC, 1925, pp. 59–79.

breathing apparatus; so before climbing down the five meters to the fallen man, the gas has to be dispersed using mechanical ventilation. The measurement of hydrogen sulfide was taken one hour after the accident; saving life, or trying to, was quite properly the first priority.

Hoidal et al. [18] have demonstrated yet another confounding factor: differences in how two people fare upon H_2S exposure, even in the same accident. They describe an accident wherein two workers entered a tank to fix a failed pump. Both rapidly lost consciousness. When emergency services extricated the workers, one of them had a much worse clinical status; he suffered severe, permanent neurological impairment (chronic vegetative state). The other worker eventually made a full recovery. The authors conclude that though both were in the same confined space, one of them was exposed to a larger amount of the toxic fumes.

4.1.3 Exposure Effects in Various Species

Table 4.1 shows the amount of H_2S at which symptoms are manifested, for various animals.

4.2 Mechanisms

Table 4.2 compares the two quite different mechanisms H_2S has for causing damage to the human body:

1. Irritant.
2. Systemic intoxication—This usually occurs at slightly higher levels. Systemic intoxication responses are more often linked to a threshold level:
   a. Cytotoxicity (interference with cellular energy production)
   b. Anoxia due to impairment of the systems controlling breathing

TABLE 4.2

Overview of Irritant and Systemic Intoxication

Mechanism	Threshold Concentration	Time	Affected Organs
Irritant	No (or very low) threshold—irritant effects begin at a few ppm	Takes longer to develop than systemic effects.	Mucous/moist membranes (eyes, nose, throat, lungs)
Systemic intoxication	Threshold level, perhaps 200 ppm	Some are extremely fast to develop, almost instantaneous.	Primarily brain and lungs; secondarily heart

At low H_2S levels, irritant effects dominate. These are usually localized to areas where mucous membranes come in contact with the ambient air, that is, the eyes, airways, and lungs. At acute poisoning levels, the clinical picture is dominated by systemic manifestations.

4.2.1 Irritant

At low to moderate H_2S concentrations, the toxic effects of H_2S are chiefly due to tissue irritation. H_2S is easily absorbed by moist membranes, such as those lining the eyes and respiratory tract. Hydrogen sulfide acts directly on these tissues to cause local inflammatory and irritative effects [20,21].

Irritant responses often take longer to develop than the systemic intoxication responses; therefore, at higher exposure levels, irritant responses may not be observed, even if they have been set in motion [20,22].

4.2.1.1 Linear, Not Threshold-Type, Response

The irritating and corrosive properties of H_2S trigger responses at very low levels, for example, 1 ppm. There seems to be a straightforward relationship between H_2S concentration and severity of the biological response.

4.2.1.2 Less Power, but Deeper Penetration into Lungs

H_2S is not nearly as powerful an irritant as chlorine, ammonia, or sulfur dioxide. However, it is capable of penetrating much deeper into the lungs, where it irritates the tissues that make up the terminal bronchial membranes [23,24].

4.2.1.3 First Symptom: "Gas Eyes"

Eye irritation is usually the first symptom reported; the H_2S acts directly on the moist membranes to cause inflammation of the conjunctiva and cornea. This is the *gas eyes* or conjunctivitis, which is well known in the petroleum industry. For more discussion of irritant effects on the eyes, see Section 4.7.

4.2.2 Hypoxia

Straddling the line between irritant effects and systemic intoxication, we have hypoxia, which can be due to one, the other, or both.

Hypoxia is an insufficient supply of oxygen to meet the demands of the body. It can occur in cases of acute H_2S exposure when the following happen:

1. H_2S induces paralysis of the brain's respiratory center (see Section 4.2.3). Breathing temporarily ceases because the diaphragm and other muscles, which cause inhalation and exhalation, simply stop. There is no incoming oxygen to the lungs.

2. Pulmonary edema interferes with oxygen exchange. There is oxygen available in the lungs, but it cannot be taken up into the blood capillaries because of the irritant damage that the H_2S has caused deep in the lung tissue (see Section 4.3.1).

In cases of acute, nonfatal H_2S intoxication, sequelae may be a result of hypoxia due to respiratory insufficiency, rather than to a direct toxic effect of hydrogen sulfide on the brain [23–27].

4.2.3 Cytotoxicity

Cytotoxicity is often cited as the mechanism that allows H_2S to render its victims unconscious with such breathtaking speed.

Cytotoxicity is the mechanism by which H_2S disrupts the principal energy-generating process at the cellular level [7,28,29].

4.2.3.1 Cell Energy Generation

Cells generate their energy when mitochondria in the cell oxidize carbon fuels. This is done through a series of energy transformations called cellular respiration.

The principal energy-generating system of the cell is oxidative phosphorylation. Electrons from energy-rich molecules are used to reduce molecular oxygen to water; in the process, a large amount of free energy is liberated. This free energy is used to generate adenosine triphosphate (ATP). ATP is the most commonly used "energy currency" of the cells in most organisms.

Oxidative phosphorylation is an extremely complex process that requires several large protein complexes to transfer electrons, and sometimes protons, through a series of reactions, to ultimately synthesize ATP [30].

4.2.3.2 Cytochrome c Oxidase

A vital component of the oxidative phosphorylation process is cytochrome c oxidase, which transports electrons. It is located in the inner membrane of the mitochondrion.

Cytochrome c oxidase (previously known as cytochrome aa3) is a multipeptide complex: it includes two related enzymes (cytochrome a and cytochrome a3), at least two copper atoms, and possibly other metals, for example, zinc and magnesium [31–33].

Cytochrome c oxidase is a key enzyme in the control of the ATP synthesis flow. It catalyzes the overall reaction:

$$c^{2+} + 4H^+ + O_2 \rightarrow 2H_2O + 4c^{3+}$$

where c^{2+} and c^{3+} are the reduced cytochrome c and oxidized cytochrome c, respectively [32].

The sequence of reactions for transferring four electrons from the reduced cytochrome c to oxygen is shown in the following equation [33]:

$$4\ \text{cyt.}c \rightarrow \text{cyt.}a \rightarrow Cu_A \rightarrow Cu_B - \text{cyt.}a_3 \rightarrow O_2$$

4.2.3.3 H₂S and Cytochrome c Oxidase

H_2S interacts with several enzymes and other macromolecules; however, the critical target enzyme is cytochrome c oxidase. H_2S binds cytochrome c oxidase and inhibits its functioning. That in turn prevents ATP generation. At high concentrations, H_2S arrests cellular respiration, depriving the cells of their energy [7,34,35].

The brain and heart are particularly sensitive to this metabolic disturbance because they have the highest oxygen requirements [17,36,37].

Nogué et al. [38] note that by interfering with cytochrome oxidase, H_2S acts as a powerful depressant for the central nervous system (CNS)—particularly of the part of the CNS that controls the body's respiratory center. Respiration is swiftly and completely paralyzed.

4.2.3.4 Slow Dissociation of the Cytochrome c Oxidase–H₂S Complex

Nicholls [39] has pointed out the cytochrome c oxidase–H₂S complex dissociates slowly. Savolainen et al. [40] have suggested that this might offer an explanation for the persistence of biochemical effects and also for the possible cumulative effects of repeated H_2S exposure.

4.2.3.5 Is Cytochrome Oxidase Disruption Responsible for "Knockdown"?

Cytochrome oxidase disruption was long thought to be the primary mechanism causing *knockdown*, also known as the "slaughterhouse sledgehammer" effect. Guidotti [1] has pointed out some problems with this assumption:

- The cells' inability to utilize oxygen, caused by the cytochrome oxidase disruption, would quickly become irreversible. The cells' ability to generate and use energy would be blocked.

- The cells' inability to utilize oxygen should be devastating to the brain and would consistently result in anoxic brain damage.

Kombian and coauthors have also questioned whether cytochrome oxidase can be responsible, in particular whether it is too slow a mechanism [41,42].

Milby and Baselt [24] have pointed out that some neurotoxic effects from H_2S exposure, such as dizziness, incoordination, and headache, appear extremely quickly—rather too quickly to be secondary effects due only to hypoxia.

The mechanism of *knockdown* should perhaps be considered an unresolved issue. It is possible that at both the high H_2S levels seen in *knockdown* and the lower levels causing dizziness and headache, there is a direct toxic effect of H_2S on the brain and CNS, which has not yet been elucidated.

4.2.3.6 Cyanide and H_2S

The action of H_2S is often compared to that of cyanide, which also blocks oxidative metabolism [20,43–46]. Some researchers, however, introduce a cautionary note about comparing H_2S to cyanide.

Beck et al. [47] have found that the effect of sulfide on bundles of frog nerves was quite different from that of cyanide. They found that the compound action potential (CAP) signal sent by the nerves was stimulated in an opposite manner for sulfide, compared to cyanide.

Truong et al. [48] have pointed out that in hydrogen cyanide poisoning, methemoglobin has been successfully used for the treatment, but that it has not been successful in treating H_2S poisonings, "even though the ferric heme group of methemoglobin scavenges H_2S."

Nicholls [39] found a difference in the chemical binding of sulfide and cyanide to cytochrome c oxidase (which he calls by its former name, cytochrome aa_3):

> Sulphide, like cyanide, is a slow-binding inhibitor of cytochrome aa_3 with a high affinity ($K_d < 0.1\ \mu M$).

> Unlike cyanide binding, the binding of sulphide is apparently independent of the redox state of components of the oxidase other than cytochrome a_3 and shows no anomalous kinetics during complex formation.

The reduced sulphide-inhibited system shows a much higher Soret peak at 445 nm than the corresponding cyanide and azide complexes, suggesting that partial electron transfer from sulphide to haem may occur in the complex. No evidence was obtained for the formation of any sulf-haem derivatives of cytochrome a_3.

4.2.4 Summary of Damage Mechanisms

If we try to summarize what is known about the damage mechanisms of H_2S, we might say

- It is well documented that H_2S inhibits cellular oxidative metabolism, which is the primary energy source for the body's cells
- H_2S inhibits the central respiratory mechanism; this is thought to be the major cause of death
- The exact mechanism of H_2S influence on the respiratory drive is not completely understood [49]

4.3 Pulmonary Effects

The cytotoxicity leading to apnea (cessation of breathing) described in Section 4.2.3 is caused by systemic intoxication. There are also pulmonary effects caused by the irritant nature of H_2S, such as

- Pulmonary edema
- Airways (bronchi and bronchiole) damage

4.3.1 Pulmonary Edema

Probably the most commonly seen pulmonary effect is edema or an accumulation of fluid in the lungs.

Pulmonary edema is usually seen at concentrations above 200 ppm. In the concentration range 200–500* ppm H_2S, a sudden onset of hemorrhagic pulmonary edema can occur, which can be fatal [14–16,50,51].

It has been suggested that the threshold-type response of edema, that is, that it kicks in at approximately 200 ppm, may be due to a sudden lowering of the surface tension of lung surfactant at this concentration [51].

---

* This "upper ceiling" is merely because above 500 ppm H_2S, other toxic effects would be expected to be rapid and lethal.

4.3.1.1 Lung Structure

When we breathe, air passes through the throat (pharynx), the voice box (larynx), and the windpipe (trachea). The trachea divides into two tubes (bronchi). These divide into smaller tubes (bronchioles). The major airways in the lungs are shown in Figure 4.1.

The bronchioles in turn divide into smaller branches called alveolar ducts. The alveolar ducts terminate in clusters of air sacs called alveoli.

The alveolus is covered by a web of blood capillaries (see Figure 4.2). The walls of the alveoli are only one cell thick. The exchange of gases between the ambient air and the body takes place here between the thin-walled alveoli and the blood vessels.

The lungs contain millions of alveoli, to allow for sufficient oxygen uptake for the large amount of energy required by mammals.

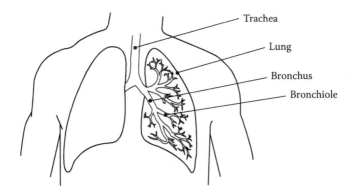

FIGURE 4.1
Major airways in the lungs.

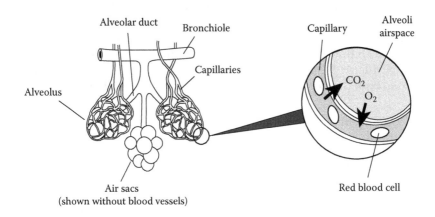

FIGURE 4.2
Alveoli and capillaries.

4.3.1.2 Fluid Buildup

In pulmonary edema, fluid collects in the lung's air sacs (alveoli). The liquid in the alveoli decreases the amount of oxygen/carbon dioxide that can be exchanged at the alveoli–blood capillary interface. The lungs' respiratory capacity is reduced and respiratory failure may result.

4.3.1.3 How Does H_2S Lead to Edema?

H_2S penetrates deep into the respiratory tract, where it causes edema through a combination of simultaneous processes [52,53]:

1. Tissue injury stimulating an inflammatory process
2. Changes in mucosal permeability
3. Reductions in ATP production

In layman's terms, it seems that at the same time as lung tissues are damaged by the corrosive character of H_2S (triggering inflammation), the larger amounts of H_2S are able to penetrate mucosal membranes, and the cells' ability to deal with it all is drastically reduced by the inhibition of ATP production (cytochrome oxidase inhibition; see Section 4.2).

4.3.1.4 Delayed Damage

It is important to note that the symptoms of pulmonary edema (buildup of fluid in the lungs), such as chest pain or shortness of breath, can be delayed for days after exposure. Shivanthan et al. [54] have described a case in Sri Lanka of delayed onset respiratory failure following an H_2S accident. After 5 days in the hospital and treatment for pneumonitis and hypoxic cardiac ischemia, the man could be discharged.

Delays between exposure and pulmonary symptoms in the literature are usually up to 72 hours; however, they can be longer. Parra et al. [55] describe a case in Spain where a patient was symptom-free for 3 weeks after the H_2S exposure, at which time dyspnea, chest tightness, and hemoptysis were reported. This particular H_2S accident had resulted in a fatality and subsequent investigation, so both the exposure date and cause were well established.

4.3.2 Damage in Larger Airways

Tanaka et al. [56] have studied the injuries of three people exposed to high concentrations of H_2S in a petroleum refinery accident in Osaka. All three survived, thanks to immediate rescue and rapid, first-class medical interventions. Serial bronchoscopy was performed for 10 days after the accident. On the third day, they reported severely injured erythematous lesion on the lingular bronchi; on the 10th day after the accident, the bronchoscopy still showed slightly injured erythematous lesion.

The authors speculate that prolonged inhalation of H_2S at lower concentrations can also produce severely injured bronchi.

4.3.2.1 Airways Reactivity

The irritant nature of H_2S would be expected to induce a reaction in the larger airways such as bronchi and bronchioles. Kennedy [57] has shown a relation between exposure to other acid gases and development of airway hyperresponsiveness.

It is one of the paradoxes of H_2S that exposure is not known to cause airways reactivity or the condition called reactive airways dysfunction syndrome (RADS). There has been a great deal of research in recent years with endogenous H_2S; and there is a possibility that endogenous amounts of H_2S act as an anti-inflammatory agent [1].

4.3.3 Pneumonitis/Pneumonia

Several reports exist of pneumonitis, aspiration pneumonia, or organizing pneumonia developing subsequent to H_2S exposure [9,54,58,59].

Parra et al. [55] have also diagnosed pneumonitis after H_2S exposure; they also reported that five months after the accident, the patient still showed exertion dyspnea and decreased lung volume, and suggest that it could be due to a mild case of interstitial fibrosis as a sequel to the H_2S exposure.

4.3.4 Chronic H$_2$S Exposure and the Lungs

4.3.4.1 Reduced Lung Capacity

Chronic low-level H_2S exposure may lead to reduced lung capacity. Richardson [60] has compared spirometric test results from 68 sewer workers to those of 60 water treatment workers (presumed to have no H_2S exposure). When comparing workers of similar age and smoking habits, a statistically significant difference was seen: the nonsmoking sewer workers achieved only 89% of the expected FEV1/FVC* values, but the nonsmoking water treatment workers achieved nearly 98%. An interesting point is that the reduction in lung function appears to be greatest in long-term sewer workers. The author suggests that annual lung function tests of sewer workers would allow for better protection of the workers.

4.3.4.2 Chronic Low-Level H$_2$S and Exercise

Bhambhani et al. [61] tested the effect of a one-time dose of 10 ppm/15 minutes on exercising adults and reported no effects on pulmonary functions. In a study

---

* Spirometry parameter: Ratio of forced expiratory volume in 1 second (FEV1), to forced vital capacity (FVC).

of exercising adults exposed to 10 ppm H_2S for 30 minutes, they saw a significant decrease in oxygen uptake and increase in blood lactate [62].

In another study, volunteers were exposed on four occasions to 1, 0.5, 2.0, or 5.0 ppm H_2S while exercising. They found that at 5.0 ppm, blood lactate levels increased, while at the lowest H_2S levels, no statistically significant results were seen, though a tendency for CO_2 output to decrease with increasing H_2S was reported [63].

Jones [64] has pointed out that the implications of this research may be that the current exposure levels are suitable for young, fit workers, but less suitable for older, less-fit workers or those with preexisting medical conditions.

4.3.5 Asthma

A concern sometimes raised in public health is the effect of low-level hydrogen sulfide exposure on asthma sufferers. For populations living near factories producing hydrogen sulfide emissions (e.g., kraft pulp mills, rayon factories, or wastewater treatment plants (WWTPs)) or in geologically active areas such as Hawaii, Japan, and New Zealand, the public health implications can be significant.

4.3.5.1 Dakota City–South Sioux City Asthma Study

Campagna et al. [65] compared daily ambient air H_2S measurements to hospital visits for asthma and respiratory diseases for the Dakota City and South Sioux City area in Nebraska.

From February 1999 through May 2000, the ATSDR* and U.S. EPA conducted extensive air monitoring of H_2S levels throughout the area. (Previous studies had shown that the communities were not exposed to levels of SO_2 at levels of concern.)

The sources of the H_2S were a slaughterhouse and leather tanning facility, with a large liquid waste treatment system, a wastewater lagoon for a truck washing facility, and a small municipal wastewater treatment plant. The largest source was the slaughterhouse's liquid waste treatment system, which was estimated to release 1900 pounds of H_2S each day.

The authors found a positive association between high H_2S levels (≥30 ppb) and following-day adult hospital visits for asthma. For children, they found a positive association between high H_2S levels and following-day hospital visits for all respiratory diseases, including asthma.

4.3.5.2 Finnish Pulp and Paper Study

Jäppinen et al. [66,67] exposed two groups of volunteers (with asthma or without asthma) to 2 ppm H_2S for 30 minutes and tested pulmonary function.

---

* U.S. Agency for Toxic Substances and Disease Registry.

They reported no effect on airway resistance in the group without asthma. It was noted, however, that in 2 of the 10 asthmatic subjects, airway resistance increased by over 30% and airway conductance decreased by over 30%, indicating airflow obstruction [66,67].

In discussing these results, Reiffenstein has pointed out that there is possibly an element of bias in the sampling: all of the volunteers were pulp mill workers. It is known that there is an element of self-selection in the pulp and paper industry: only those who can tolerate the fumes work in this environment [6,27].

4.3.5.3 Iraqi Oil Field Workers Study

Mousa [68] has investigated chronic low-level H_2S exposure among Iraqi oil field workers. The study included one oil field with ambient H_2S levels up to 50 ppm and three oil fields with little or no ambient H_2S as controls.

The most frequent symptoms following ambient H_2S exposure were nasal bleeding (53% of patients), pharyngeal bleeding, and gum and mouth bleeding. Other less frequent symptoms included headache, abdominal colic, and fatigue.

No correlation was seen between asthma attacks and ambient H_2S exposure. However, this may be due to the same self-selection process as in the pulp and paper industry: the author points out that those with asthma problems tend to not work in oil fields.

4.3.5.4 What Can We Conclude?

We are convinced that these studies are yielding important information about the health implications of living or working in these particular environments. In general, however, there are two points that recur in the discussions of these studies:

1. Copollutants, both chemical and solid particles. The industries of interest, as well as geothermal areas, can emit other, equally damaging chemicals, for example, NO_x and SO_x. Fine particulate matter may also occur in geologically active areas or pulp and paper mills. These copollutants are expected to play a significant role in pulmonary effects.
2. There is always the possibility of an element of "self-selection" in these studies. People who suffer effects from low-level H_2S may tend to remove themselves from the scene by changing jobs.

For these reasons, we draw a rather cautious conclusion: for those who have asthma, low levels of H_2S in the air may lead to increased airway obstruction. We would welcome more studies in this area.

4.4 Neurological Sequelae

The systemic effects of H_2S intoxication, especially the cessation of breathing, are often immediate and life-threatening; they are dramatic and frankly tend to steal the show. The area of neurological effects, we feel, deserves more attention than it receives today. Though neurological effects are often less immediate and dramatic, they can cause huge amounts of long-term suffering to the victims of H_2S exposure and their families.

Neurological effects described in the literature include [3,4,37,55,58,69–73]

- Ataxia
- Convulsions
- Dizziness
- Headache
- Nausea
- Persistent vegetative state
- Prolonged coma
- Pseudobulbar or cognitive impairment
- Spasticity
- Tremor
- Vertigo
- Vomiting
- Weakness
- Weight loss

4.4.1 Neurological Effects in Exposed versus Unexposed Populations

Symptoms that are often reported in the literature, such as headaches or short-term memory loss, are very nonspecific; they may occur as a response to causes other than H_2S. It is very useful, therefore, to have studies of neurological effects in both control (unexposed) populations and groups exposed to H_2S.

Kilburn [74,75] has described long-term (2–20+ years) neurobehavioral effects of 19 subjects after H_2S exposure (which ranged in duration from "moments" to years). Their physiological and psychological test results were compared to those of an unexposed population.

Physiology measurements showed that those exposed to H_2S had impaired balance, prolonged simple and choice reaction times, decreased visual fields, abnormal color discrimination, and decreased hearing. Psychological domains showed cognitive disability, reduced perceptual motor speed, impaired verbal recall and remote memory, and abnormal mood status among those exposed to H_2S.

Interestingly, Kilburn reports similar results between those whose exposure included unconsciousness and those who did not undergo loss of consciousness and were therefore presumably exposed to a lesser amount of H_2S [74,75].

Similar results were reported in another study of people working at or living downwind from a sour crude oil processing plant versus a control population unexposed to H_2S [76], and again in a New Mexican study of three communities exposed to H_2S from a sewage plant, sour gas installations, and the waste lagoons of a cheese-manufacturing plant, compared to a similar, but unexposed, population in Arizona [77].

In evaluating this group of studies, the *self-selection* concerns discussed in Section 4.3.5.4 also apply here. The alert reader will also have noted that these studies stem from one research group. We should like to see other research groups performing such studies of neurological effects; we believe that this will make it easier to interpret the overall results.

4.4.2 Typical Cases

4.4.2.1 Kilburn: Worker Rendered Semiconscious by H₂S

A typical case report is given by Kilburn. An oil well tester, 23 years of age, was exposed to H_2S; the level was unknown but he was rendered semiconscious by it. He received oxygen; the following day, he was hospitalized because symptoms of nausea, vomiting, diarrhea, and incontinence had developed. Thirty-nine months after the accident, a physical exam, chest x-rays, and pulmonary function tests were normal (though vibration sense was diminished). However, the neurological tests showed impaired balance, delayed two-choice reaction times, impaired recall, and impaired cognitive functions. Scores for confusion, tension–anxiety, depression, and fatigue were elevated. Further testing 49 months after the accident showed some improvements but not a return to normalcy [78].

4.4.2.2 Policastro and Otten: Acute H₂S Exposure in Confined Space

Policastro and Otten have vividly described the aftereffects of another accident, where two workers experienced acute, brief exposure to H_2S in a confined space [79]:

> Patient #1 was extubated on day 2 post-injury with no focal neurological dysfunction. He had no cranial nerve dysfunction, sensory, or motor paralysis. The patient experienced psychiatric symptoms including acute anxiety, grief reaction, and depression. He developed transient thrombocytopenia, which was felt to be dilutional. There was some transient chest pain, but it was ruled-out for acute myocardial infarction. Patient #1 was discharged from the hospital on day 5 post-exposure, and

psychiatric follow-up was recommended. He continues to suffer from anxiety and depression since the event.

Patient #2 remained in a persistent vegetative state with severe cerebral edema, multiorgan system dysfunction, and coagulopathy despite aggressive resuscitative measures. Ultimately, care was withdrawn and he expired on day 4 post-exposure.

4.4.2.3 Schneider et al.: Sewer Construction Worker Rendered Unconscious by H_2S

Schneider et al. [21] have provided us with a glimpse of what these neurological effects actually mean in a person's life. They describe a fatal H_2S accident in New Jersey, involving the construction crew laying the foundation for a municipal sewage pumping station. (The same accident is described by Snyder et al.; see Section 5.7.1.) In this incident, a worker was overcome by H_2S in a 27-foot-deep pit; another worker went down to attempt to rescue the first; and a police officer who then tried to rescue the two workers was fatally overcome by the H_2S.

Of the two workers who survived, one was treated with hyperbaric oxygen and released from hospital 48 hours later. His coworker, who had entered the pit to attempt a rescue, suffered prolonged illness due to neurological and pulmonary injury. Instead of listing symptoms, we quote the authors' descriptions of the patient before and after the H_2S accident:

> *Before*: The patient was a previously healthy 27-year-old male construction worker. Throughout grade school, he was a popular, well-adjusted, conscientious student who performed at or above grade level on achievement/academic indices. He attended and graduated from a vocational high school with a major in auto mechanics. He received excellent or outstanding grades in categories such as achievement, punctuality, ambition, learning experience and quality/quantity of work and graduated with a 3.0 grade point average. After graduation he worked on construction-related jobs and was well-known locally for his work as a designer and builder of Monster Trucks.

> *After*: The picture that emerges from neuropsychological testing is of an individual with a variety of problems such as bradyphrenia, poor memory and impaired planning ability. However, this list of deficits fails to convey the actual impact of this toxic exposure on this patient's life. Observations by the psychiatrist he has been seeing since the accident are necessary to complete the portrait. "He frequently, in this interview, deferred judgement issues to his wife for her decisions because he realizes he is not able to adequately function independently. He shows overt contradiction, sees the overt contradiction in his comments, and yet cannot integrate the contradictory thinking. For example, he indicates to me that it made sense for the state to take his license because he is unable to react appropriately as he drives; however, it makes sense

for him to drive when they return his license. He sees the contradiction but does nothing with it; he just shrugs his shoulders and says he is fine. He does not report any independent activities that are sustained in any manner, shape or form. He describes essentially a totally goal-less existence." [21]

4.4.3 Physical Evidence Provided by CT and MRI

The impact of H_2S exposure on the human brain has been shown in computed tomography (CT) images. Gaitonde et al. [80] used computed tomograms in the treatment of a 20-month-old child who had been exposed for approximately one year to low levels of H_2S from the burning tip of a colliery near the family's home. The precise amount of H_2S exposure is unknown but was at least 0.6 ppm in the home. On admission to the hospital, the child had "gross truncal ataxia, choreoathetosis, and dystonia and could not stand." A computed tomogram showed "striking bilateral areas of low attenuation in the basal ganglia and in some of the surrounding white matter." A repeat scan, taken 10 weeks after the child's exposure to H_2S ceased, showed complete resolution of the abnormalities of the basal ganglia.

Matsuo et al. [72] used CT to study the brain of the victim of a fatal industrial H_2S accident. They found similar results: low-density areas in the basal ganglia.

Nam et al. [81] performed magnetic resonance imaging (MRI) scans as part of the diagnosis and treatment of a 25-year-old man exposed to H_2S in a refinery accident. The MRI findings were similar to cortical laminar necrosis represented in hypoxic brain damage.

The lack of evidence of damage on CT or MRI images does not mean that such damage to the brain has not occurred. Tvedt et al. [58] describe a patient suffering from acute H_2S poisoning from a Norwegian shipyard accident. The neurological sequelae were severe enough that, five years afterward, the man was unable to resume work. CT scans after four months and an MRI scan after five years showed slight atrophy but no changes in white matter or globus pallidus.

4.4.4 Dosage and Exposure Time

Tvedt et al. [37] have emphasized the importance of two variables: dose and exposure time. They found that exposure time is a key predictor of the amount of long-term neurological damage from H_2S intoxication.

Those who spend less than one minute in the H_2S atmosphere, even if they lose consciousness, tend to fare better than those who were unconscious in the H_2S atmosphere for more than five minutes. The neurological symptoms included—but were not limited to—amnesic syndromes (transitory or permanent; slight to serious); motor symptoms such as ataxia, tremor, and rigidity; and increased susceptibility to strong odors. One patient suffered serious dementia, which lasted until his death 8 years after the accident.

4.4.5 Needed: More Data, More Follow-Up

It seems plausible that the permanent neurological sequelae are historically underreported, for a number of reasons:

- The changes to the CNS caused by H_2S, though permanent, can be subtle.
- Neuropsychological and neurological testing are simply not part of the traditional long-term follow-up. Historically, after an H_2S incident, the focus is more on tracking pulmonary or cardiopulmonary effects.

More information on chronic neurological sequelae is needed. The limited data makes it difficult to correlate the risks of sequelae to the H_2S concentration and exposure time.

4.4.5.1 Recommendations for Follow-Up

Annual neurological and neuropsychological testing for at least 5 years is recommended by some researchers, for patients who show evidence of neurotoxicity after H_2S exposure, because of the potential chronic neurological sequelae [3,59].

4.5 Myocardial Effects

The heart is almost certainly not a primary target organ for H_2S toxicity, since cardiac function continues through the pulmonary and CNS disruptions [23,82]. But even if H_2S does not directly act on the myocardium, it can indirectly have a high impact. Cardiac muscle has a high oxygen demand. When H_2S interferes with oxygen demand, for example, through pulmonary edema or through CNS effects, the cardiac muscle can suffer [27,36].

Myocardial effects reported in the literature include the following [3,4,83,84]:

- Secondary electrocardiograph changes
- Myocardial infarction
- Convulsions and tachycardia with evidence of myocardial damage
- Dilated cardiomyopathy and, upon hospital release 23 days after the accident, decreased cardiac function [84]

At low concentrations, the indications seem to be that H_2S does not present a risk to the cardiovascular system [63].

4.6 Immune System

An area that deserves more attention is the effect of H_2S on the body's ability to fight off disease. There are a number of reports in the medical literature of patients developing pneumonia or bronchopneumonia as a sequel to H_2S poisoning [4,9,54,59].

It seems fairly clear that there is a relationship between H_2S exposure and an impaired intrapulmonary antibacterial defense system, though the mechanism is perhaps imperfectly understood.

4.6.1 Impairment of Alveolar Macrophages

The alveolar regions of the lungs are protected against bacterial infection by a system of phagocytes, also known as macrophages. Macrophages are cells derived from the bone marrow, with the purpose of rapidly intercepting and ingesting foreign bacteria [85]. The macrophages in the lungs form the alveolar macrophage (AM) system. The AM mechanism is a highly complex defense system; broadly speaking, it involves the following:

1. Phagocytes are attracted to the bacterium by chemotactic factors. Serum opsonins attach to the bacterial surface. These act almost as "seasoning"—phagocytes that are surrounded by many palatable particles or cells will choose to ingest the ones that have been opsonized.
2. The phagocyte ingests the bacterial cell.
3. The phagocyte employs a complex range of microbicidal enzymes and toxic substances to inactivate the bacterial cell.

The phagocytic mechanism is complex and can be disrupted at a number of places by a toxin such as ozone, NO_x, or H_2S. Chemotactic substances can be destroyed, so that the AM do not select the invading bacteria for ingestion. Phagocytic mobility might be reduced, reducing the ability to rapidly intercept invading bacteria. The AM may be structurally damaged, or its enzyme systems impaired [86,87].

There have been a number of studies examining the effect of H_2S on the immune system of the lungs, especially the AMs. These findings are summarized in Table 4.3.

Rogers and Ferin [92] have shown that rats who breathed low amounts of H_2S (45 ppm for 2, 4, or 6 hours) have a decreased ability for the lungs to subsequently inactivate *Staphylococcus epidermidis*. The mechanism was

TABLE 4.3

Animal Studies Examining Impairment of Alveolar Macrophages after H_2S Exposure

| Subject | Study | H_2S Exposure | | Effects | Reference |
		Level (ppm)	Duration (hours)		
Rabbit	In vitro, cultured AMs	50 and 200	8–48	Dramatic increase in cell damage (measured by phagocytosis and trypan blue exclusion).	[88]
Rabbit	In vitro, cultured AMs	50–60	24	Macrophage cells have 95% reduction in phagocytic activity.	[89]
Rabbit	In vivo, biomarkers	100–200	1	Elevated concentrations of thiosulfate in blood and urine.	[90]
Guinea pigs	In vivo, lung cells lavaged after exposure	200	24	16% decrease in macrophage number; 6% decrease in phagocytic ability.	[91]
Rats	In vivo, lung cells lavaged after exposure	200	24	50% decrease in macrophage number; 26% decrease in phagocytic ability.	[91]
Rats	In vivo, lungs	45	2, 4 or 6	Decreased ability for the lungs to inactivate *Staphylococcus epidermidis*; believed to be due to the effect of H_2S on AMs.	[92]
Rats	In vivo, lung cells lavaged after exposure	0, 50, 200, 400	4	Significant decrease in pulmonary AM viability seen at 400 ppm.	[93]

not definitively established, but the authors proposed that the reduced ability to inactivate staphylococcus is due to H_2S-induced impairment of AM, since, among other things, staphylococci normally are rapidly phagocytized. The authors suggest that this might help explain secondary pneumonias in humans after acute or subacute H_2S exposure.

The effect of H_2S on the number of AM and their phagocytic ability has been studied by Robinson and coworkers. Robinson [88,89] examined the effect of H_2S on rabbit AM cultured on gas-permeable membranes. After 24 hours of 60 ppm H_2S exposure, the cultured macrophage cells exhibited a 95% reduction in phagocytic activity. Transmission electron microscopy of the H_2S-exposed cells showed cytoplasmic membrane degradation and disruption of interior organelles. The control cultures, not exposed to H_2S, did not have a reduction in phagocytic activity. The H_2S exposure in these experiments was in vitro exposure of cultured cells.

It seems reasonable to expect that the same pattern would be seen using in vivo rabbit experiments. Kage et al. [90] have demonstrated that when rabbits are exposed in vivo to H_2S at similar levels to those used by Robinson, the H_2S biomarkers appear in the blood—so it therefore was absorbed in the lungs.

Robinson et al. [91] conducted in vivo H_2S exposures on guinea pigs and rats. After exposure, the lungs were lavaged and the exfoliated cells were studied for type, quantity, viability, and, in the case of AM, phagocytic ability. For guinea pigs, the H_2S exposure resulted in somewhat worsened AM number and activity; for rats it was considerably more.

Working with rats, Khan et al. [93] saw a significant decrease in pulmonary AM viability. This decrease occurred at 400 ppm, but not at 50 or 200 ppm. The difference in results may simply reflect the different levels of H_2S tolerated by the various types of animals (see Section 4.1.3).

4.6.2 Decreased Cytochrome Oxidase Activity

Dorman et al. [94] exposed rats to nontoxic levels (up to 400 ppm) of H_2S for 3 hours and measured the levels of sulfide and thiosulfate, as well as cytochrome oxidase activity, in lung, liver, nose, and hindbrain tissue before, during, and after the H_2S exposure. Rats were exposed to the H_2S, either once or five times.

Decreased cytochrome oxidase activity was seen in the olfactory epithelium after 3 hours at 30 ppm H_2S exposure (see Table 4.4). Decreased cytochrome oxidase activity was not observed in the hindbrain. In a related study of chronic exposure [95], rats were exposed to lower levels for 6 hours/day for 70 days. After the 70-day exposure at 80 ppm H_2S, cytochrome oxidase activity in the lung tissue decreased from 1.04 to 0.87 U/mg protein—a statistically significant decrease.

For the olfactory epithelium samples, the falloff in cytochrome oxidase activity is dramatically steeper after five exposures. This data is shown in Figure 4.3. This may imply that there is a cumulative effect taking place.

TABLE 4.4

Mean Hindbrain and Nasal Epithelium Cytochrome Oxidase Activity, Following One or Five 3-hour H₂S Exposures

	Cytochrome Oxidase	
H₂S Exposure	Single 3-hour Exposure	Five 3-hour Exposures
Hindbrain		
0	2.14 ± 0.12	Not determined
200	2.55 ± 0.26	Not determined
400	2.28 ± 0.31	Not determined
Respiratory epithelium		
0	1.23 ± 0.05	1.02 ± 0.17
30	0.60 ± 0.08[a]	0.83 ± 0.11
80	0.50 ± 0.08[a]	0.60 ± 0.19
200	0.92 ± 0.02[a]	0.74 ± 0.07
400	0.94 ± 0.02[a]	0.86 ± 0.19
Olfactory epithelium		
0	1.20 ± 0.06	1.13 ± 0.09
30	1.01 ± 0.07[a]	0.75 ± 0.13[a]
80	0.99 ± 0.07[a]	0.84 ± 0.11[a]
200	0.92 ± 0.03[a]	0.70 ± 0.12[a]
400	0.92 ± 0.04[a]	0.51 ± 0.01[a]

Source: Data from Dorman, D.C. et al., *Toxicol. Sci.*, 65(1), 18, 2002.
Note: H₂S exposure is given in ppm, while cytochrome oxidase in U/mg protein.
[a]   Statistically different from control values ($p \leq 0.05$).

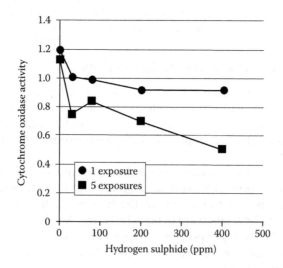

FIGURE 4.3
Cytochrome oxidase activity in rat olfactory epithelium following H₂S exposure. (Data from Dorman, D.C. et al., *Toxicol. Sci.*, 65(1), 18, 2002.)

4.6.3 Important Questions

The studies described earlier have laid a very important groundwork, and we should like to see more research building on this. There are still important questions to be answered:

1. Is the AM impairment the cause of the secondary infections, for example, pneumonia, associated with H_2S exposures? It seems reasonable, if not proven. It is known that a decreased rate of bacteria-killing relates directly to decreased pulmonary resistance to infection [86].

2. What is the mechanism involved in AM impairment?
   a. The work of Robinson indicates that structural damage to the macrophage occurs, which may imply irritant effects.
   b. On the other hand, the work of Dorman et al. seems to implicate cytotoxicity.
   c. An interference with chemotactic substances, or the AM enzyme system, may also be responsible.

3. How do animal models correspond to human disease rates?

4. Are the effects of H_2S on the lungs' defense system cumulative effects, or delayed sequelae, or is there a threshold below which the impairment will not occur?

The fourth question has important implications both in treatment of victims after an H_2S incident and in deciding appropriate monitoring programs and limits for chronic low-level exposures.

It may be that the full expression of H_2S on the lungs' ability to fight microorganisms is a combination of cumulative *and* delayed effects. Also, it may be that there is a threshold of concentration x time that must be reached before the damage occurs. More research is needed to answer these and other questions.

4.7 Effects on Cornea and Conjunctiva

The eye's cornea is an ideal target tissue for H_2S, since the gas is water soluble. The moist mucous membranes of the eyes are directly exposed to H_2S in the air. Starting at low concentrations, the cornea and conjunctiva become irritated and attain a condition known as "conjunctivitis."

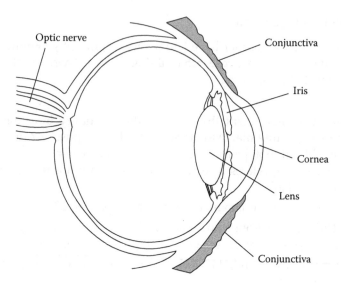

FIGURE 4.4
Cross section of the eye.

4.7.1 Conjunctivitis

Conjunctivitis, or "pink eye," is the most common eye effect of H_2S exposure [96]. It is an inflammation or infection of the conjunctiva, the membrane that lines the eyelids and protects exposed parts of the sclera or white part of the eyeball. The conjunctiva is shown in Figure 4.4.

   Conjunctivitis is a painful condition that can range from mild to severe. In mild cases, the eyes itch and smart; the lids feel dry and rough. As the condition worsens, lacrimation and pus-like secretions develop, and the victim experiences pain caused by light (photophobia). In severe cases, vision may be permanently impaired due to ulcers and scarring on the cornea.

4.7.1.1 Symptoms of Conjunctivitis

The symptoms of conjunctivitis include the following [97–103]:

- Soreness, smarting, or itching
- Sensation of grains of sand in the eyes
- Haloes or rainbow effects around artificial lights
- Lacrimation
- Photophobia
- Blepharospasm

- Superficial punctate corneal erosions
- Cornea covered with pinpoint inflamed blisters over the area least protected by the eyelids

4.7.1.2 Delay in Symptoms

It has often been noted that the symptoms of "gas eyes" may not appear for some hours after exposure; pain and vision problems often occur in the evening after a work shift [19,99,101,102]. Sometimes the symptoms are most severe the day following the exposure.

Larsen [101] has suggested that the slight anesthetic effect of the H_2S reduces symptoms during exposure and is responsible for the delayed effect.

4.7.1.3 Other Names for Conjunctivitis

Conjunctivitis is often referred to as "pink eye" or "sore eye"; in the petroleum and gas industry, it is also known as "gas eyes" [98,104]. In the H_2S literature it is sometimes also called keratoconjunctivitis [103].

4.7.2 Animal Studies of H_2S and the Eyes

Some of the animal toxicity studies regarding the effect of H_2S on the eyes are summarized in Table 4.5.

4.7.2.1 Exfoliative Cytology Experiments

Exfoliative cytology is an interesting technique that may offer an objective measure of ocular irritation. After exposure to H_2S, the eyes are washed

TABLE 4.5

Animal Studies: Effect of H_2S on Eyes

Subject	H_2S Level (ppm)	Duration	Effects	Reference
Rats	54	3 hours	Lesions of the cornea.	[105]
Calves	20, 150	7 days	Photophobia and corneal opacity.	[106]
Rats	35–65	4–8 hours	Irritation of eyes.	[19]
		18–48 hours	Pus in eyes.	
Rats	400	4 hours	Number of exfoliated conjunctival and corneal epithelial cells was circa twice that of unexposed rats.	[104]
	1500	4 minutes	Number of exfoliated conjunctival and corneal epithelial cells was circa twice that of unexposed rats.	
Dogs	100	4–8 hours	Lacrimation.	[19]
		8–16 hours	Pus in eyes.	

TABLE 4.6

Exfoliated Epithelial Cells in Lavage Fluid

H_2S Level	Duration	Exfoliated Epithelial Cells in Lavage Fluid
0	4 hours	19 cells/μL
400 ppm	4 hours	44 cells/μL
1500 ppm	4 minutes	35 cells/μL

Source: Data from Lefebvre, M. et al., *Vet. Hum. Toxicol.*, 33(6), 564, 1991.

(lavaged) with saline solution. The lavage fluid is collected, and the number of epithelial cells in it is counted. Lefebvre et al. [104] used this method on rats exposed to 0 or 400 ppm H_2S for 4 hours, or 1500 ppm for 4 minutes. They found that with H_2S exposure, the quantity of exfoliated cells was increased (see Table 4.6) and the proportion of conjunctival to corneal epithelial cells increased [27,104,107].

4.8 Other Effects

4.8.1 H_2S and Spontaneous Abortions

There are not so many studies on this, possibly because the attention to H_2S historically has focused on fatalities—which are overwhelmingly male. The industries most associated with H_2S—wastewater, petrochemical, and pulp and paper—tend to have a predominantly male workforce. But there are some studies with mixed results.

Hemminki and Niemi [108] examined miscarriages among women working in the Finnish rayon and pulp and paper industries or whose husbands worked in these industries. The pollutants of concern were sulfur dioxide, hydrogen sulfide, and carbon disulfide. Their study included measurements of H_2S in the ambient air. They found that in areas where the mean annual level of H_2S was greater than 4 μg/m^3, more spontaneous abortions occurred in all socioeconomic classes—but they also note that this result is not statistically significant.

Xu et al. [109] studied miscarriages among 2800 women at a petrochemical complex in Beijing and found a statistically significant correlation between H_2S exposure and rate of spontaneous abortions. It must be noted, however, that the H_2S exposure information is not based on air sampling but was self-reported; it is based on interviews and depends on the women's perceptions of the level of exposure. It is not clear if the reported H_2S exposures are truly H_2S or possibly include other odiferous sulfur-containing compounds.

Petrochemical plants can be expected to deal with a number of sulfur compounds, including H_2S, SO_2, and carbon disulfide (CS_2), and it is possible that these can confound results. Saillenfait et al. [110] have examined the effects on pregnant rats of inhaling CS_2 or combinations of CS_2 and H_2S. They found that CS_2 leads to an increase in fetal toxicity and birth defects and that this is increased by adding H_2S to the CS_2.

4.8.2 H₂S and Cancer

No correlation has been found between ambient H_2S and cancer [111,112]. We should like to point what the U.S. EPA states:

> This substance (H_2S) has not undergone a complete evaluation and determination under US EPA's Integrated Risk Information System (IRIS) program for evidence of human carcinogenic potential. [113]

Nor has H_2S been specifically evaluated by the International Agency for Research on Cancer [114].

References

1. Guidotti, T. L. (2010). Hydrogen sulfide advances in understanding human toxicity. *International Journal of Toxicology*, 29(6), 569–581.
2. Voight, G. E. and Mueller, P. (1955). Versuche sum histochemischen Nachweis der Schwefelwasserstoff-Vergiftung (The histochemical effect of hydrogen sulfide poisoning). *Acta Histochemistry*, 1, 223–239.
3. Snyder, J. W., Safir, E. F., Summerville, G. P., and Middleberg, R. A. (1995). Occupational fatality and persistent neurological sequelae after mass exposure to hydrogen sulfide. *The American Journal of Emergency Medicine*, 13(2), 199–203.
4. Kemper, F. D. (1966). A near-fatal case of hydrogen sulfide poisoning. *Canadian Medical Association Journal*, 94(21), 1130.
5. Arnold, I. M., Dufresne, R. M., Alleyne, B. C., and Stuart, P. J. (1985). Health implication of occupational exposures to hydrogen sulfide. *Journal of Occupational and Environmental Medicine*, 27(5), 373–376.
6. Ferris, Jr., B. G. (1978). Health effects of exposure to low levels of regulated air pollutants: A critical review. *Journal of the Air Pollution Control Association*, 28(5), 482–497.
7. Guidotti, T. L. (1994). Occupational exposure to hydrogen sulfide in the sour gas industry: Some unresolved issues. *International Archives of Occupational and Environmental Health*, 66(3), 153–160.
8. Durand, M. and Weinstein, P. (2007). Thiosulfate in human urine following minor exposure to hydrogen sulfide: Implications for forensic analysis of poisoning. *Forensic Toxicology*, 25(2), 92–95.

9. Knight, L. D. and Presnell, S. E. (2005). Death by sewer gas: Case report of a double fatality and review of the literature. *The American Journal of Forensic Medicine and Pathology*, 26(2), 181–185.

10. Deprez, R. D., Oliver, C., and Halteman, W. (1986). Variations in respiratory disease morbidity among pulp and paper mill town residents. *Journal of Occupational and Environmental Medicine*, 28(7), 486–491.

11. Jaakkola, J. J., Paunio, M., Virtanen, M., and Heinonen, O. P. (1991). Low-level air pollution and upper respiratory infections in children. *American Journal of Public Health*, 81(8), 1060–1063.

12. Partti-Pellinen, K., Marttila, O., Vilkka, V., Jaakkola, J. J., Jäppinen, P., and Haahtela, T. (1996). The South Karelia air pollution study: Effects of low-level exposure to malodorous sulfur compounds on symptoms. *Archives of Environmental Health: An International Journal*, 51(4), 315–320.

13. Finnbjornsdottir, R. G., Oudin, A., Elvarsson, B. T., Gislason, T., and Rafnsson, V. (2015). Hydrogen sulfide and traffic-related air pollutants in association with increased mortality: A case-crossover study in Reykjavik, Iceland. *BMJ Open*, 5(4), e007272.

14. Lopez, A., Prior, M., Yong, S., Albassam, M., and Lillie, L. E. (1987). Biochemical and cytologic alterations in the respiratory tract of rats exposed for 4 hours to hydrogen sulfide. *Toxicological Sciences*, 9(4), 753–762.

15. Prior, M. G., Sharma, A. K., Yong, S., and Lopez, A. (1988). Concentration-time interactions in hydrogen sulphide toxicity in rats. *Canadian Journal of Veterinary Research*, 52(3), 375.

16. Prior, M., Green, F., Lopez, A., Balu, A., DeSanctis, G. T., and Fick, G. (1990). Capsaicin pretreatment modifies hydrogen sulphide-induced pulmonary injury in rats. *Toxicologic Pathology*, 18(2), 279–288.

17. Christia-Lotter, A., Bartoli, C., Piercecchi-Marti, M. D., Demory, D., Pelissier-Alicot, A. L., Sanvoisin, A., and Leonetti, G. (2007). Fatal occupational inhalation of hydrogen sulfide. *Forensic Science International*, 169(2), 206–209.

18. Hoidal, C. R., Hall, A. H., Robinson, M. D., Kulig, K., and Rumack, B. H. (1986). Hydrogen sulfide poisoning from toxic inhalations of roofing asphalt fumes. *Annals of Emergency Medicine*, 15(7), 826–830.

19. Sayers, R. R., Smith, N. A. C, Fieldner, A. C., Mitchell, C. W., Jones, G. W., Yant, W. P., Stark, D. D., Katz, S. H., Bloomfield, J. J., and Jacobs, W. A. (1925). Investigation of toxic gases from Mexican and other high-sulphur petroleums and products. *US Bureau of Mines, Bulletin 231*. Government Printing Office: Washington, DC, pp. 59–79.

20. Beauchamp, R. O., Bus, J. S., Popp, J. A., Boreiko, C. J., Andjelkovich, D. A., and Leber, P. (1984). A critical review of the literature on hydrogen sulfide toxicity. *CRC Critical Reviews in Toxicology*, 13(1), 25–97.

21. Schneider, J. S., Tobe, E. H., Mozley, P. D., Barniskis, L., and Lidsky, T. I. (1998). Persistent cognitive and motor deficits following acute hydrogen sulphide poisoning. *Occupational Medicine*, 48(4), 255–260.

22. Kleinfeld, M., Giel, C., and Rosso, A. (1964). Acute hydrogen sulphide intoxication: An unusual source of exposure. *Industrial Medicine and Surgery*, 33, 656.

23. Haggard, H. W. (1925). The toxicology of hydrogen sulfide. *Journal of Industrial Hygiene*, 7, 113.

24. Milby, T. H. and Baselt, R. C. (1999). Hydrogen sulfide poisoning: Clarification of some controversial issues. *American Journal of Industrial Medicine*, 35:192–195.

25. Gabbay, D. S., De Roos, F., and Perrone, J. (2001). Twenty-foot fall averts fatality from massive hydrogen sulfide exposure. *The Journal of Emergency Medicine*, 20(2), 141–144.

26. Milby, T. H. (1962). Hydrogen Sulfide Intoxication: Review of the Literature and Report of Unusual Accident Resulting in Two Cases of Nonfatal Poisoning. *Journal of Occupational and Environmental Medicine*, 4(8), 431–437.

27. Reiffenstein, R. J., Hulbert, W. C., and Roth, S. H. (1992). Toxicology of hydrogen sulfide. *Annual Review of Pharmacology and Toxicology*, 32(1), 109–134.

28. Khan, A. A., Schuler, M. M., Prior, M. G., Yong, S., Coppock, R. W., Florence, L. Z., and Lillie, L. E. (1990). Effects of hydrogen sulfide exposure on lung mitochondrial respiratory chain enzymes in rats. *Toxicology and Applied Pharmacology*, 103(3), 482–490.

29. Nicholls, P. and Kim, J. K. (1982). Sulphide as an inhibitor and electron donor for the cytochrome c oxidase system. *Canadian Journal of Biochemistry*, 60(6), 613–623.

30. Berg, J. M., Tymoczko, J. L., and Stryler, L. (2002). Chapter 18: Oxidative phosphorylation, in *Biochemistry*, 5th ed. W. H. Freeman: New York.

31. Wainio, W. W. (1983). Cytochrome c oxidase: TWO models. *Biological Reviews*, 58(1), 131–156.

32. Hill, B. C., Greenwood, C., and Nicholls, P. (1986). Intermediate steps in the reaction of cytochrome oxidase with molecular oxygen. *Biochimica et Biophysica Acta (BBA)—Reviews on Bioenergetics*, 853(2), 91–113.

33. Manon, S., Camougrand, N., and Guerin, M. (1989). Inhibition of the phosphate-stimulated cytochrome c oxidase activity by thiophosphate. *Journal of Bioenergetics and Biomembranes*, 21(3), 387–401.

34. Nicholson, R. A., Roth, S. H., Zhang, A., Zheng, J., Brookes, J., Skrajny, B., and Bennington, R. (1998). Inhibition of respiratory and bioenergetic mechanisms by hydrogen sulfide in mammalian brain. *Journal of Toxicology and Environmental Health Part A*, 54(6), 491–507.

35. Fujita, Y., Fujino, Y., Onodera, M., Kikuchi, S., Kikkawa, T., Inoue, Y., Niitsu, H., Takahashi, K., and Endo, S. (2011). A fatal case of acute hydrogen sulfide poisoning caused by hydrogen sulfide: Hydroxocobalamin therapy for acute hydrogen sulfide poisoning. *Journal of Analytical Toxicology*, 35(2), 119–123.

36. Ammann, H. M. (1986). A new look at physiologic respiratory response to H2S poisoning. *Journal of Hazardous Materials*, 13(3), 369–374.

37. Tvedt, B., Skyberg, K., Aaserud, O., Hobbesland, A., and Mathiesen, T. (1991). Brain damage caused by hydrogen sulfide: A follow-up study of six patients. *American Journal of Industrial Medicine*, 20(1), 91–101.

38. Nogué, S., Pou, R., Fernández, J., and Sanz-Gallén, P. (2011). Fatal hydrogen sulphide poisoning in unconfined spaces. *Occupational Medicine*, 61(3), 212–214.

39. Nicholls, P. (1975). The effect of sulphide on cytochrome aa3 Isosteric and allosteric shifts of the reduced α-peak. *Biochimica et Biophysica Acta (BBA)—Bioenergetics*, 396(1), 24–35.

40. Savolainen, H., Tenhunen, R., Elovaara, E., and Tossavainen, A. (1980). Cumulative biochemical effects of repeated subclinical hydrogen sulfide intoxication in mouse brain. *International Archives of Occupational and Environmental Health*, 46(1), 87–92.

41. Kombian, S. B., Warenycia, M. W., Mele, F. G., and Reiffenstein, R. J. (1987). Effects of acute intoxication with hydrogen sulfide on central amino acid transmitter systems. *Neurotoxicology*, 9(4), 587–595.

42. Kombian, S. B., Reiffenstein, R. J., and Colmers, W. F. (1993). The actions of hydrogen sulfide on dorsal raphe serotonergic neurons in vitro. *Journal of Neurophysiology*, 70(1), 81–96.

43. Smith, R. P. (1986). Toxic responses of the blood, in *Casarett and Doull's Toxicology—The Basic Science of Poisons*, 3rd ed., eds. C. D. Klaassen, M. O. Amdur, and J. Doull. Macmillan: New York, pp. 223–244.

44. Gossel, T. A. and Bricker, J. D. (1994). *Principles of Clinical Toxicology*, 3rd ed. Raven Press: New York, pp. 109–134.

45. Chaturvedi, A. K., Smith, D. R., and Canfield, D. V. (2001). A fatality caused by accidental production of hydrogen sulfide. *Forensic Science International*, 123(2), 211–214.

46. Wang, R. (2012). Physiological implications of hydrogen sulfide: A whiff exploration that blossomed. *Physiological Reviews*, 92(2), 791–896.

47. Beck, J. F., Donini, J. C., and Maneckjee, A. (1983). The influence of sulfide and cyanide on axonal function. *Toxicology*, 26(1), 37–45.

48. Truong, D. H., Eghbal, M. A., Hindmarsh, W., Roth, S. H., and O'Brien, P. J. (2006). Molecular mechanisms of hydrogen sulfide toxicity. *Drug Metabolism Reviews*, 38(4), 733–744.

49. Svendsen, K. (2001). The Nordic expert group for criteria documentation of health risks from chemicals and The Dutch Expert Committee on Occupational Standards: 127. Hydrogen sulphide. Stockholm, Sweden.

50. Burnett, W. W., King, E. G., Grace, M., and Hall, W. F. (1977). Hydrogen sulfide poisoning: Review of 5 years' experience. *Canadian Medical Association Journal*, 117(11), 1277.

51. Green, F. H., Schurch, S., De Sanctis, G. T., Wallace, J. A., Cheng, S., and Prior, M. (1991). Effects of hydrogen sulfide exposure on surface properties of lung surfactant. *Journal of Applied Physiology*, 70(5), 1943–1949.

52. Holman, R. G. and Maier, R. V. (1990). Oxidant-induced endothelial leak correlates with decreased cellular energy levels. *American Review of Respiratory Disease*, 141(1), 134–140.

53. Roth, S. H. and Goodwin, V. M. (2003). *Health Effects of Hydrogen Sulphide: Knowledge Gaps*. Alberta Environment: Edmonton, Alberta, Canada.

54. Shivanthan, M. C., Perera, H., Jayasinghe, S., Karunanayake, P., Chang, T., Ruwanpathirana, S., Jayasinghe, N., De Silva, Y., and Jayaweerabandara, D. (2013). Hydrogen sulphide inhalational toxicity at a petroleum refinery in Sri Lanka: A case series of seven survivors following an industrial accident and a brief review of medical literature. *Journal of Occupation Medical Toxicology*, 8(1), 9.

55. Parra, O., Monsó, E., Gallego, M., and Morera, J. (1991). Inhalation of hydrogen sulphide: A case of subacute manifestations and long term sequelae. *British Journal of Industrial Medicine*, 48(4), 286.

56. Tanaka, S., Fujimoto, S., Tamagaki, Y., Wakayama, K., Shimada, K., and Yoshikawa, J. (1999). Bronchial injury and pulmonary edema caused by hydrogen sulfide poisoning. *The American Journal of Emergency Medicine*, 17(4), 427–429.

57. Kennedy, S. M. (1992). Acquired airway hyperresponsiveness from nonimmunogenic irritant exposure. *Occupational Medicine*, 7(2), 287–300.

58. Tvedt, B., Edland, A., Skyberg, K., and Forberg, O. (1991). Delayed neuropsychiatric sequelae after acute hydrogen sulfide poisoning: Affection of motor function, memory, vision and hearing. *Acta Neurologica Scandinavica*, 84(4), 348–351.

59. Doujaiji, B. and Al-Tawfiq, J. A. (2010). Hydrogen sulfide exposure in an adult male. *Annals of Saudi Medicine*, 30(1), 76.
60. Richardson, D. B. (1995). Respiratory effects of chronic hydrogen sulfide exposure. *American Journal of Industrial Medicine*, 28(1), 99–108.
61. Bhambhani, Y., Burnham, R., Snydmiller, G., MacLean, I., and Lovlin, R. (1996). Effects of 10-ppm hydrogen sulfide inhalation on pulmonary function in healthy men and women. *Journal of Occupational and Environmental Medicine*, 38(10), 1012–1017.
62. Bhambhani, Y., Burnham, R., Snydmiller, G., and MacLean, I. (1997). Effects of 10-ppm hydrogen sulfide inhalation in exercising men and women: Cardiovascular, metabolic, and biochemical responses. *Journal of Occupational and Environmental Medicine*, 39(2), 122–129.
63. Bhambhani, Y. and Singh, M. (1991). Physiological effects of hydrogen sulfide inhalation during exercise in healthy men. *Journal of Applied Physiology*, 71(5), 1872–1877.
64. Jones, K. (2014). Case studies of hydrogen sulphide occupational exposure incidents in the UK. *Toxicology Letters*, 231(3), 374–377.
65. Campagna, D., Kathman, S. J., Pierson, R., Inserra, S. G., Phifer, B. L., Middleton, D. C., Zarus, G. M., and White, M. C. (2004). Ambient hydrogen sulfide, total reduced sulfur, and hospital visits for respiratory diseases in northeast Nebraska, 1998–2000. *Journal of Exposure Science and Environmental Epidemiology*, 14(2), 180–187.
66. Jäppinen, P., Vilkka, V., Marttila, O., and Haahtela, T. (1990). Exposure to hydrogen sulphide and respiratory function. *British Journal of Industrial Medicine*, 47(12), 824–828.
67. Jäppinen, P. and Tenhunen, R. (1990). Hydrogen sulphide poisoning: Blood sulphide concentration and changes in haem metabolism. *British Journal of Industrial Medicine*, 47(4), 283.
68. Mousa, H. A. L. (2015). Short-term effects of subchronic low-level hydrogen sulfide exposure on oil field workers. *Environmental Health and Preventive Medicine*, 20(1), 12–17.
69. Wasch, H. H., Estrin, W. J., Yip, P., Bowler, R., and Cone, J. E. (1989). Prolongation of the P-300 latency associated with hydrogen sulfide exposure. *Archives of Neurology*, 46(8), 902–904.
70. Ellenhorn, M. J., and Barceloux, D. G. (1988). Hydrogen sulfide, in *Medical Toxicology: Diagnosis and Treatment of Human Poisoning*, eds. M. J. Ellenhorn and D. G. Barceloux. Elsevier: New York, pp. 836–840.
71. Berlin, C. M. (1981). Death from anoxia in an abandoned cesspool. *Annals of Internal Medicine*, 95(3), 387–387.
72. Matsuo, F., Cummins, J. W., and Anderson, R. E. (1979). Neurological sequelae of massive hydrogen sulfide inhalation. *Archives of Neurology*, 36(7), 451–452.
73. Hurwitz, L. J. and Taylor, G. (1954). Poisoning by sewer gas: With unusual sequelae. *The Lancet*, 263(6822), 1110–1112.
74. Kilburn, K. H. (1997). Exposure to reduced sulfur gases impairs neurobehavioral function. *Southern Medical Journal*, 90(10), 997–1006.
75. Kilburn, K. H. (2003). Effects of hydrogen sulfide on neurobehavioral function. *Southern Medical Journal*, 96(7), 639–646.

76. Kilburn, K. H. and Warshaw, R. H. (1995). Hydrogen sulfide and reduced-sulfur gases adversely affect neurophysiological functions. *Toxicology and Industrial Health*, 11(2), 185–197.
77. Kilburn, K. H., Thrasher, J. D., and Gray, M. R. (2010). Low-level hydrogen sulfide and central nervous system dysfunction. *Toxicology and Industrial Health*, 26(7), 387–405.
78. Kilburn, K. H. (1993). Case report: Profound neurobehavioral deficits in an oil field worker overcome by hydrogen sulfide. *The American Journal of the Medical Sciences*, 306(5), 301–305.
79. Policastro, M. A. and Otten, E. J. (2007). Case files of the University of Cincinnati fellowship in medical toxicology: Two patients with acute lethal occupational exposure to hydrogen sulfide. *Journal of Medical Toxicology*, 3(2), 73–81.
80. Gaitonde, U. B, Sellar, R. J., and O'Hare, A. E. (March 7, 1987). Long term exposure to hydrogen sulphide producing subacute encephalopathy in a child. *British Medical Journal*, 294, 614.
81. Nam, B., Kim, H., Choi, Y., Lee, H., Hong, E. S., Park, J.-K., Lee, K.-M., and Kim, Y. (2004). Neurologic sequela of hydrogen sulfide poisoning. *Industrial Health*, 42(1), 83–87.
82. Almeida, A. F. and Guidotti, T. L. (1999). Differential sensitivity of lung and brain to sulfide exposure: A peripheral mechanism for apnea. *Toxicological Sciences*, 50(2), 287–293.
83. Gregorakos, L., Dimopoulos, G., Liberi, S., and Antipas, G. (1995). Hydrogen sulfide poisoning: Management and complications. *Angiology*, 46(12), 1123–1131.
84. Lee, E. C., Kwan, J., Leem, J. H., Park, S. G., Kim, H. C., Lee, D. H., Kim, J. H. and Kim, D. H. (2009). Hydrogen sulfide intoxication with dilated cardiomyopathy. *Journal of Occupational Health*, 51(6), 522–525.
85. Phair, J. P. (1980). Infection and resistance: A review, in *Wastewater Aerosols and Disease: Proceeding of a Symposium*, eds. H. Pahren and W. Jakubowski. Environmental Protection Agency, Health Effects Research Laboratory: Cincinnati, OH.
86. Goldstein, E., Jordan, G W., and MacKenzie, M. R. (1976). Methods for evaluating the toxicological effects of gaseous and particulate contaminants on pulmonary microbial defense systems. *Annual Review of Pharmacology and Toxicology*, 16: 447–463.
87. Pierce, C. W. (1980). Macrophages: Modulators of immunity. Parke-Davis Award Lecture. *The American Journal of Pathology*, 98(1), 10.
88. Robinson, A. V. (1980). Effect of in vitro exposure to hydrogen sulfide on cultured rabbit alveolar macrophages. In: *Pacific Northwest Laboratory Annual Report for 1979 to the DOE Assistant Secretary for Environment, Part 1, Biomedical Sciences*. Pacific Northwest Laboratory, Report PNL-3300, pp. 279–281. Available from National Technical Information Service (NTIS): Springfield, VA.
89. Robinson, A. V. (1982). Effect of in vitro exposure to hydrogen sulfide on rabbit alveolar macrophages cultured on gas-permeable membranes. *Environmental Research*, 27(2), 491–500.
90. Kage, S., Nagata, T., Kimura, K., Kudo, K., and Imamura, T. (1992). Usefulness of thiosulfate as an indicator of hydrogen sulfide poisoning in forensic toxicological examination: A study with animal experiments. *Japan Journal of Forensic Toxicology*, 10(3), 223–227.

91. Robinson, A. V., McDonald, K. E., and Renne, R. A. (1980). Effect of in vivo exposure to hydrogen sulfide on free cells obtained from lungs of rats and guinea pigs. In: *Pacific Northwest Laboratory Annual Report for 1979 to the DOE Assistant Secretary for Environment, Part 1, Biomedical Sciences.* Pacific Northwest Laboratory, Report PNL-3300, pp. 282–283. Available from National Technical Information Service (NTIS): Springfield, VA.

92. Rogers, R. E. and Ferin, J. (1981). Effect of hydrogen sulfide on bacterial inactivation in the rat lung. *Archives of Environmental Health: An International Journal,* 36(5), 261–264.

93. Khan, A. A., Yong, S., Prior, M. G., and Lillie, L. E. (1991). Cytotoxic effects of hydrogen sulfide on pulmonary alveolar macrophages in rats. *Journal of Toxicology and Environmental Health, Part A Current Issues,* 33(1), 57–64.

94. Dorman, D. C., Moulin, F. J. M., McManus, B. E., Mahle, K. C., James, R. A., and Struve, M. F. (2002). Cytochrome oxidase inhibition induced by acute hydrogen sulfide inhalation: Correlation with tissue sulfide concentrations in the rat brain, liver, lung, and nasal epithelium. *Toxicological Sciences,* 65(1), 18–25.

95. Dorman, D. C., Brenneman, K. A., Struve, M. F., Miller, K. L., James, R. A., Marshall, M. W., and Foster, P. M. (2000). Fertility and developmental neurotoxicity effects of inhaled hydrogen sulfide in Sprague–Dawley rats. *Neurotoxicology and Teratology,* 22(1), 71–84.

96. Nyman, H. T. (1954). Hydrogen sulfide eye inflammation: Treatment with cortisone. *Industrial Medicine & Surgery,* 23, 161.

97. Sayers, R. R., Mitchell, C. W., and Yant, W. P. (1923). Hydrogen sulphide as an industrial poison. Reports of Investigations, Serial No. 2491. US Bureau of Mines, Department of the Interior: Washington, DC.

98. Yant, W. P. (June 1930). Hydrogen sulphide in industry—Occurrence, effects, and treatment. *American Journal of Public Health and the Nations Health,* 20(6), 598–608.

99. Sjögren, H. (1939). A contribution to our knowledge of the ocular changes induced by sulphuretted hydrogen. *Acta Ophthalmologica,* 17(2), 166–171.

100. Howes, H. S. (1944). Eye inflammation as the only symptom of incipient hydrogen sulphide poisoning. *Analyst,* 69, 92.

101. Larsen, V. (1944). Eye diseases caused by hydrogen sulphide in tunnel workers. *Acta Ophthalmologica,* 21(4), 271–286.

102. Beasley, R. W. R. (1963). The eye and hydrogen sulphide. *British Journal of Industrial Medicine,* 20(1), 32–34.

103. Luck, J. and Kaye, S. B. (1989). An unrecognised form of hydrogen sulphide keratoconjunctivitis. *British Journal of Industrial Medicine,* 46(10), 748.

104. Lefebvre, M., Yee, D., Fritz, D., and Prior, M. G. (1991). Objective measures of ocular irritation as a consequence of hydrogen sulphide exposure. *Veterinary and Human Toxicology,* 33(6), 564–566.

105. Michal, F. V. (1950). Eye lesions caused by hydrogen sulfide. *Ceskoslovenska Oftalmologie,* 6, 5–8.

106. Nordstrom, G. A. and McQuitty, J. B. (1975). Response of calves to atmospheric hydrogen sulfide and ammonia. Paper No. 75–212. *Proceedings of Canadian Society of Agricultural Engineering,* 75(212).

107. NRC. (2010). *Acute Exposure Guideline Levels for Selected Airborne Chemicals: Volume 9.* National Research Council (USA), Committee on Acute Exposure Guideline Levels. National Academies Press: Washington, DC.

108. Hemminki, K. and Niemi, M. L. (1982). Community study of spontaneous abortions: Relation to occupation and air pollution by sulfur dioxide, hydrogen sulfide, and carbon disulfide. *International Archives of Occupational and Environmental Health*, 51(1), 55–63.

109. Xu, X. et al. (1998). Association of petrochemical exposure with spontaneous abortion. *Occupational and Environmental Medicine*, 55(1), 31–36.

110. Saillenfait, A. M., Bonnet, P., and De Ceaurriz, J. (1989). Effects of inhalation exposure to carbon disulfide and its combination with hydrogen sulfide on embryonal and fetal development in rats. *Toxicology Letters*, 48(1), 57–66.

111. Bates, M. N., Garrett, N., Graham, B., and Read, D. (1998). Cancer incidence, morbidity and geothermal air pollution in Rotorua, New Zealand. *International Journal of Epidemiology*, 27(1), 10–14.

112. ATSDR. (2006). Toxicological profile for hydrogen sulfide. U.S. Department of Health and Human Services, Public Health Service, Agency for Toxic Substances and Disease Registry (ATSDR): Atlanta, GA. Publication July 2006 [online]. Available: http://www.atsdr.cdc.gov/toxprofiles/tp114.pdf. Accessed April 2, 2010.

113. EPA. (1998). Record for H2S. Integrated Risk Information System (IRIS), Environmental Protection Agency (USA): Washington, DC. Available at: http://www.epa.gov/iris/subst/0061.htm. Accessed December 10, 2015.

114. IARC. (2009). Overall evaluations of carcinogenicity to humans, in *IARC Monographs on the Evaluation of Carcinogenic Risks to Humans*. International Agency for Research on Cancer (IARC): Lyon, France.

5

Biomarkers for Hydrogen Sulfide Poisoning

5.1 Introduction

Sensors and agents that would allow accurate, fast detection of hydrogen sulfide have long been desired by physicians and other workers in this field. A method that would allow real-time, or near-real-time, determination of H_2S intoxication could have many applications, for example

- Aid in emergency room diagnosing
- Establish cause of death
- Warn of chronic low-level H_2S exposure

The most commonly studied biomarkers are

- Sulfide in blood
- Thiosulfate in blood
- Thiosulfate in urine

Thiosulfate is a major metabolite of hydrogen sulfide. In healthy nonexposed people, thiosulfate occurs in very low concentrations in both the blood and urine. Finding it can be taken as a strong indication of H_2S poisoning.

In recent years, a great deal of investigation has been done as to whether sulfide or thiosulfate levels in the blood, or thiosulfate levels in urine, can be used as biomarkers, if only retrospectively. The results are summarized in Section 5.2 and discussed in Sections 5.3 and 5.4.

5.1.1 Lack of Clinical Tests as Diagnostic Aid

Methods for detecting/measuring H_2S in the body have been developed using colorimetry, electrochemistry, and gas chromatography. However, at least at the present time, none of them are sufficiently rapid for clinical diagnosis.

In the scientific literature, there is a palpable sense of frustration from emergency room physicians that rapid H_2S tests are simply not available to aid in diagnosis. Several authors—for example, Gabbay et al. [1], Yalamanchili and Smith [2], Jones [3], Peng et al. [4], and Milby and Baselt [5]—have remarked on the lack of rapid H_2S blood tests for clinical diagnosis. As Milby and Baselt write, "The delay in obtaining a result from a reference laboratory precludes incorporating blood levels into the diagnostic process."

Urine thiosulfate is sometimes proposed as a diagnostic aid in cases of nonfatal H_2S intoxication [6]. However, others have noted that while urine thiosulfate levels can be determined, the results are not available in time to aid in critical situations [1–5,7].

5.1.2 Establishing H_2S as Cause of Death

It can be difficult to establish H_2S as the definitive cause of death at autopsy. Knight and Presnell [8] investigated a fatal accident in South Carolina involving two sewer workers. Postmortem toxicology results were not clear-cut, probably due to postmortem metabolism. Determination of cause of death relied heavily on descriptions of eyewitnesses and first responders; laboratory findings were consistent with H_2S intoxication but not *specific* to it.

Christia-Lotter et al. [9] report a similar experience: in a fatality in France involving a sewer worker: the cause of death was evident, but direct toxicological proof was not obtained from the blood tests. They noted that if the context of the accident is not known, then "the gross lesions of H_2S intoxication are not specific: diffuse congestion of the internal organs, petechial haemorrhage of the internal organs, which is observed in all asphyxia syndromes, and serous and haemorrhagic pulmonary edema."

5.1.2.1 Variations within the Same Accident

Ago et al. [10] investigated a case of H_2S poisoning in the ballast tank of a cargo ship. Ten workers in the ballast tank presented symptoms of H_2S poisoning, and two died. The autopsy found high levels of thiosulfate in the blood of both victims; but there was a marked difference in the erosion of the respiratory tract and petechiae of the mouth's mucous membrane between the two men. The authors suggest that they may have been exposed to different levels of H_2S and that the respiration of one victim possibly arrested earlier than the other.

The case points out that the pathological and toxicological findings of H_2S can be expected to vary not only between incidents but within the victims of the same incident.

5.1.2.2 Absence of Biomarkers

It should be emphasized that the absence of these biomarkers in blood, tissue, or urine does not preclude H_2S intoxication as the cause of death.

As Policastro and Otten [11] put it, "No single readily-available biological diagnostic test reliably confirms exposure to H_2S in a clinically efficient manner. Indirect laboratory findings, however, may be suggestive of hydrogen sulfide toxicity."

We will give the last word on this to two doctors from the Occupational Safety and Health Administration who recommend:

> H_2S poisoning should be considered if a victim suffers a sudden collapse, especially in a confined space such as a sewer, well, or septic tank. A "rotten egg" smell may be reported at the scene or detected upon autopsy. Other autopsy finding may include hemorrhagic pulmonary edema, visceral congestion, scattered petechiae, myocarditis, or a greenish color of the brain, viscera, and bronchial secretions. Pathologic findings are variable. [12]

5.1.3 Other Fluids or Tissues

Tominaga et al. [13] have suggested that pericardial fluid (PCF) and bone marrow aspirate (BMA) may be used as an alternative to blood analysis for postmortem toxicological analysis. They point out that PCF and BMA are well-preserved postmortem materials in many H_2S intoxication cases, and these substances can be easily collected in substantial amounts at postmortem. In the case they studied, substantially lower concentrations were found in PCF than in blood and BMA.

Maebashi et al. [14] have suggested that cerebral spinal fluid may be used for detecting sulfide or thiosulfate or both.

Dorman et al. [15] have pointed out the following:

- Free sulfide is volatile and is expected to be lost from the tissue sample; special care must be taken to minimize the loss.
- Different types of tissue show very different behavior in absorbing sulfide when H_2S exposure occurs. Details are given in Section 5.4.1.2.

There is very little data on PCF, BMA, or cerebral spinal fluid analyses in cases of H_2S poisonings; we are convinced that a great deal of work needs to be done before this routinely yields useful results.

5.2 Data Found in Literature

As mentioned in Section 5.1.1, over the past several decades, there have been various methods developed for measuring H_2S and its metabolites in the body, based on colorimetry, electrochemistry, and gas chromatography. In this section, we limit the data presented and analyzed to that based on chromatography methods.

5.2.1 Normal Values

Baseline levels reported in the literature, that is, levels of endogenous sulfide or thiosulfate in people who have not been exposed to H_2S, are varied and depend on the procedures used to extract the sulfide or thiosulfate from the tissue and the analytical methods used to quantify it. Values found in the literature are shown in Table 5.1.

5.2.1.1 Urine Thiosulfate Variations

Setting a baseline value for the normal (unexposed) population can be confounded by diet; ingesting food or drink containing sulfur—for example, asparagus—can dramatically increase urinary thiosulfate levels. Chwatko and Bald [24] measured thiosulfate levels in urine collected at regular intervals throughout the days from 13 healthy adults (6 men and 7 women). The first two samples were taken before breakfast and the last at 10:30 p.m. They found a very sharp peak after breakfast, and a smaller peak in the evening, as shown in Figure 5.1.

5.2.1.2 Suggested Upper Value for General Unexposed Population

Jones [3] has suggested that for the general unexposed population, a reasonable upper value for thiosulfate in urine of 7.9 mmol/mol creatinine might be used. This is Kangas and Savolainen's mean plus two standard deviations.

TABLE 5.1

Reported Sulfide or Thiosulfate Levels in People Who Have Not Been Exposed to H_2S

Substance	Amount	Reference
Normal body content of inorganic sulfides	0.05 mg/L	[16]
Sulfide in blood	<0.05 µg/mL	[17]
	0.03 µg/mL	[18]
Thiosulfate in blood	<0.003 µmol/mL (<0.336 µg/mL)	[19,20]
	0.03 µg/mL	[11]
	0.3 µg/mL	[12]
Thiosulfate in urine	0.031 µmol/day (28.2 nmol/mL)	[20]
	0.4–5.4 mmol/mol creatinine (avg. 2.9 ± 2.5)	[21]
	≤8 µg/mL	[11]
	7.2 µmol/mol creatinine	[22]
	25.1 µmol/mg creatinine	[23]
	1.36–4.89 mmol/mol creatinine (avg. 2.85 ± 0.94)	[24]

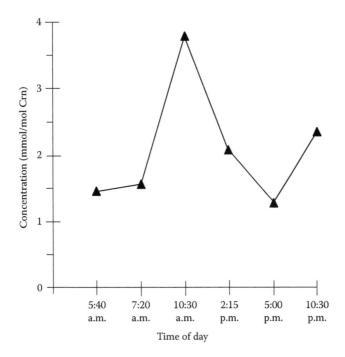

FIGURE 5.1
Urinary thiosulfate levels throughout the day, normalized on creatinine. (Reprinted from Chwatko, G. and Bald, E., *Talanta*, 79(2), 229, 2009. With permission.)

5.2.2 Reported Values, Accidents

Table 5.2 shows blood and urine biomarker data found in the literature for two geothermal, one domestic, and several industrial H_2S accidents.

5.2.3 Reported Values, Suicides

In these cases, H_2S is intentionally generated and inhaled for the express purpose of killing oneself (see Section 5.2.3.1). Table 5.3 shows the biomarker data found in the literature for H_2S suicide victims. These cases are treated separately from the accident data, because the concentrations of H_2S inhaled are expected to be much higher than in most industrial accidents.

5.2.3.1 H₂S Suicide Fad

Until recently, almost all cases of fatal H_2S poisoning were industrial accidents. We have found no references in the literature to anyone deliberately manufacturing H_2S gas with a view to self-destruction, before September 2006 [38].

TABLE 5.2W

H$_2$S Detection Using Blood and Urine Sampling

Incident	Description	Code	Notes	Sulfide in Blood (µg/mL)	Thiosulfate in Blood (µg/mL)	Thiosulfate in Urine (µg/mL)	Reference
1.	One person; fatal	C		0.305	n.r.	n.r.	[25]
2.	Nine people; one fatality	C	Person 2A (recovery in 48 hours)	N.D. (serum)	1.3 (serum)	n.r.	[26]
			Person 2B (prolonged illness)	N.D. (serum)	2.0 (serum)	n.r.	
			Person 2C (dead on scene)	N.D. (serum)	12.0 (serum)	n.r.	
			Six more injured people	N.D. (serum)	N.D. (serum)	n.r.	
3.	Four people; no fatalities	C	Person 3A	N.D.	N.D.	48.22	[6]
			Person 3B	N.D.	N.D.	43.73	
			Person 3C	N.D.	N.D.	13.46	
			Person 3D	N.D.	N.D.	N.D.	
4.	One person; fatal	C		0.224	2.80	n.r.	[6]
5.	One person; fatal	C	Four hours after death	0.958	11.21	n.r.	[6]
			At autopsy	30.34	11.21	n.r.	
6.	One person; fatal	C	(Geothermal power plant, oil separator room)	n.r.	48× higher than normal pop.	n.r.	[27]
7.	One person; fatal	C		1.68	n.r.	n.r.	[18]
8.	Four people; three fatalities	C	Person 8A (dead on scene)	0.13	10.53	0.90	[28]
			Person 8B (dead on scene)	0.11	4.59	N.D.	
			Person 8C (died 22 days later)	N.D.	4.14 (plasma)	137.20	
			Person 8D (survived)	N.D.	N.D.	29.34	

(Continued)

TABLE 5.2 (*Continued*)

H₂S Detection Using Blood and Urine Sampling

Incident	Description	Code	Notes	Sulfide in Blood (μg/mL)	Thiosulfate in Blood (μg/mL)	Thiosulfate in Urine (μg/mL)	Reference
9.	Four people; all fatalities	C		6–187× higher than normal pop.	37–77× higher than normal pop.		[29]
10.	One nonfatality	C	16-year-old male (recovered in 2 weeks)	n.r.	n.r.	107	[30]
11.	Two people; both fatalities	C	Person 11A	n.r.	0.09	n.r.	[8]
			Person 11B	n.r.	0.09	n.r.	
12.	Two fatalities	C	Person 12A	n.r.	9.98	n.r.	[10]
			Person 12B	n.r.	15.922	n.r.	
13.	One fatality	C	Blood sample obtained antemortem	n.r.	3.5	n.r.	[2]
			Blood sample obtained postmortem	n.r.	N.D.	n.r.	
14.	One fatality	G		0.68	23.547	"Normal"	[31]
15.	Two people; both fatalities	G	Person 15A	n.r.	5	2	[32]
			Person 15B	n.r.	14	8	
16.	Three people; all fatalities	C	Person 16A (dead on scene)	2.48	n.r.	n.r.	[33]
			Person 16B (dead on scene)	14.9	n.r.	n.r.	
			Person 16C (dead on scene)	18.1	n.r.	n.r.	

(*Continued*)

TABLE 5.2 (*Continued*)

H$_2$S Detection Using Blood and Urine Sampling

Incident	Description	Code	Notes	Sulfide in Blood (µg/mL)	Thiosulfate in Blood (µg/mL)	Thiosulfate in Urine (µg/mL)	Reference
17.	Two people; no fatalities	C	Person 17A (lost consciousness for 2 days; fourteen days in hospital)	0.479	8.073	229.87	[34]
			Person 17B (transient loss of consciousness; ten days in hospital)	0.447	2.691	30.275	
18.	Two people; both fatalities	D	26-year-old woman; H$_2$S concentration in lung tissue: 1.46 mg/kg	N.D.	n.r.	n.r.	[35]
			Nine-month-old daughter; H$_2$S concentration in lung tissue: 1.92 mg/kg	N.D.	n.r.	n.r.	
19.	Two people; no fatalities	C	Person 19A	n.r.	n.r.	36.6	[3]
			Person 19B	n.r.	n.r.	1.12	
20.	Two people; no fatalities; rendered unconscious by exposure	C	Person 20A	n.r.	N.D.	n.r.	[3]
			Person 20B	n.r.	N.D.	n.r.	
21.	One fatality	C		n.r.	2.47	N.D.	[3]

Code: G, geothermal incident; C, industrial accident; D, domestic accident. N.D., not detected; n.r., not reported or not measured.

TABLE 5.3

H₂S Detection Using Blood and Urine Sampling

Case	Description	Code	Notes	Sulfide in Blood (µg/mL)	Thiosulfate in Blood (µg/mL)	Thiosulfate in Urine (µg/mL)	Reference
1.	One person; fatal	S		0.66	15.698	"Normal"	[36]
2.	One person; fatal	S		0.22	38.124	n.r.	[37]
3.	One person; fatal	S		8.589	13.007	2.243	[34]
4.	One fatality	S		0.22	n.r.	n.r.	[14]
5.	Two fatalities	S		4.07	10.54	1.794	[14]
		**	** = relative of a suicide, accidently killed by fumes; outside of room where suicide took place	0.76	1.57	N.D.	[14]
6.	One fatality	S		5.63	11.55	0.448	[14]
7.	One fatality	S		0.20	3.028	299.27	[14]
8.	One fatality	S		30.90	72.66	6.28	[14]
9.	One fatality	S		2.35	40.37	N.D.	[14]
10.	One fatality	S		1.33	N.D.	n.r.	[14]
11.	One fatality	S		0.17	2.243	n.r.	[14]
12.	One fatality	S		3.24	11.213	n.r.	[14]
13.	One fatality	S		10.20	4.485	7.288	[14]
14.	One fatality	S		31.84	39.133	N.D.	[14]
15.	One fatality	S		0.44	2.018	40.703	[14]
16.	One fatality	S		0.37	1.57	n.r.	[14]
17.	One fatality	S		0.11	3.92	4.709	[14]

Code: S, bath salt suicide; N.D., not detected; n.r., not reported or not measured.

Most tragically, this changed when someone posted a new suicide method on a Japanese Internet message board: mix two household products—a popular bath essence plus toilet detergent—in a confined space; breathe in the H_2S gas generated. Victims in most cases lose consciousness in a single breath and die more or less immediately.

The fad caught on quickly after a highly publicized incident in 2008. In Japan the spring of 2008 was the worst. Between March 27 and June 15, 2008, 220 cases of attempted H_2S gas suicides were reported, killing 208. Some were women, but the overwhelming majority were men, with the age group 20–29-year-olds dominating (followed by 30–39-year-olds).

There were secondary disasters, too. This method produces H_2S gas in huge quantities, far more than that needed to kill one person (and usually much higher than is seen in industrial accidents). Suicides took place in residential bathrooms, in hotel rooms, in tents, in cars, and even in sealed-up plastic bags. The great quantities of lethal gas generated in small spaces caused disasters for rescuers and family members. At least one parent of a suicide victim was also killed in a secondary disaster [14]. In another tragedy, the husband of a suicide victim was also killed [39]. The Tokyo Fire Department warned families, hotel staff, and neighbors to not enter any rooms where H_2S was suspected to have been made. Also, "second-hand H_2S" sometimes caused injuries to paramedics and caregivers [40,41].

In Japan the "H_2S suicide craze" was 2007–2009, with 2008 being the worst. The craze has since been reported in the United States (2008, 2 cases; 2009, 10 cases; 2010, 18 cases) and Great Britain, though not nearly as severe; and recently at least one such suicide has occurred in Australia [42–45].

5.3 Boxplot Analyses of Data Found in the Literature

Box-and-whisker diagrams (boxplots) were created for the 54 cases in Tables 5.2 and 5.3 for which there were numerical values.

5.3.1 Blood Values

The boxplot for blood sulfide is shown in Figure 5.2. The three groups are nonfatal H_2S accidents, fatal H_2S accidents, and fatal H_2S suicides. The fatalities were depicted as two groups because the amounts of H_2S ingested in suicides are expected to be extremely high compared to normal industrial accidents.

The boxplots for blood thiosulfate are shown in Figure 5.3. The three groups are the same as earlier: nonfatal H_2S accidents, fatal H_2S accidents, and fatal H_2S suicides.

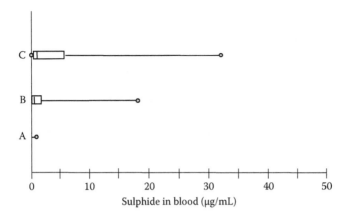

FIGURE 5.2
Boxplots of blood sulfide values. A: nonfatal accidents (n = 9). B: fatal accidents (n =15). C: fatal suicides (n = 17).

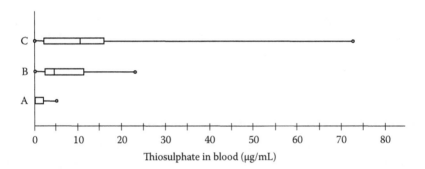

FIGURE 5.3
Boxplots of blood sulfide values. A: nonfatal accidents (n = 11). B: fatal accidents (n =16). C: fatal suicides (n = 16).

5.3.2 Urine Values

The boxplot for urine thiosulfate is shown in Figure 5.4. The two groups are nonfatal H_2S accidents and fatal H_2S accidents/suicides.

5.3.3 What Do These Tell Us?

The three boxplots indicate the following:

1. In cases where H_2S intoxication is suspected, *the absence of sulfide and/or thiosulfate in blood and urine samples cannot eliminate H_2S as the suspected causative agent.* There are too many cases in the figures and tables described earlier where the biomarker levels are zero, but the person was killed nevertheless.

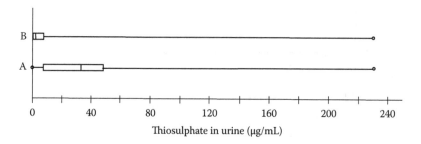

FIGURE 5.4
Boxplot of urine thiosulfate values. A: nonfatal accidents ($n = 10$). B: Fatal accidents and suicides ($n = 16$).

2. For survivors of an H_2S accident, urine samples *might* yield useful information.

3. In fatal cases, blood samples may help corroborate H_2S as cause of death.

4. Both blood thiosulfate and blood sulfide should be analyzed in the case of fatalities. It is not possible to know, until the analyses are done, which will predominate.

5.4 Biomarker Complications and Limitations

Sulfide and thiosulfate levels in the body are not static: they are constantly changing as the compounds metabolize and are processed through the body. Sulfide in the blood quickly metabolizes to thiosulfate; this eventually shows up in the urine and then is excreted from the body (see Figure 5.5).

Death is a surprisingly complex variable in this. Death does not stop the metabolism of H_2S in the blood; but it does stop the bodily functions that produce urine. Thus, if death occurs very quickly after exposure, thiosulfate never has time to reach the urine [37]. If, however, an appreciable amount of time elapses between H_2S exposure and death, then the blood sulfide and

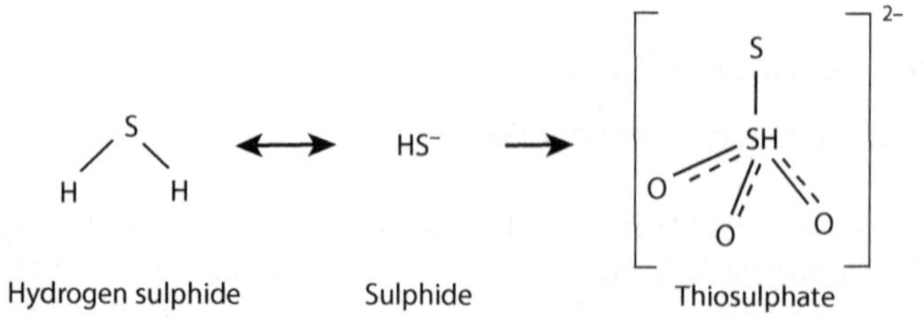

Hydrogen sulphide   Sulphide   Thiosulphate

FIGURE 5.5
Hydrogen sulfide metabolism.

thiosulfate levels fall with time (see Section 5.4.3.2), and thiosulfate may appear in the urine. Once death actually occurs, the picture changes again, as decomposition processes in blood and tissue start producing hydrogen sulfide.

5.4.1 Relationship between H_2S Air Levels and Biomarker Levels

It should be emphasized that quantitative relationships between air concentration of H_2S and levels of sulfide/thiosulfate in blood, urine, or tissue have not been established.

Nor has there been established a correlation between duration of exposure and biomarker levels.

There is not much data on this in the literature; we feel this is a field that would benefit from more research.

5.4.1.1 Work of Dorman et al. with Rats Exposed to H_2S

As noted earlier, there is no direct way that ppm of H_2S in the atmosphere can be converted to H_2S levels in tissue. One interesting study that does exist is the work of Dorman et al. [15] with rats, which has been described in Section 4.6.2. They exposed rats to nontoxic levels (up to 400 ppm) of H_2S for 3 hours and measured the levels of sulfide and thiosulfate, as well as cytochrome oxidase activity, in the lung, liver, nose, and hindbrain tissue before, during, and after the H_2S exposure. Some rats were exposed to the H_2S five times.

During the 3-hour exposure, an increase was seen in the sulfide concentrations of the lung (from 0.54 to 0.88 µg/g), liver, and nasal tissue, but not in the hindbrain tissue. Sulfide results for the nose and hindbrain are given in Table 5.4.

TABLE 5.4

Mean Hindbrain and Nasal Epithelium Sulfide Concentrations Immediately after the End of a Single 3-hour H_2S Exposure

Tissue	H_2S Exposure (ppm)	Sulfide Concentration (µg/g) (Single Exposure)
Hindbrain	0	1.21 ± 0.05
	200	1.12 ± 0.05
	400	1.14 ± 0.04
Respiratory epithelium	0	1.73 ± 0.14
	200	1.37 ± 0.11
	400	2.73 ± 0.77
Olfactory epithelium	0	1.42 ± 0.11
	200	1.25 ± 0.06
	400	2.07 ± 0.33[a]

Source: Data from Dorman, D.C. et al., *Toxicol. Sci.*, 65(1), 18, 2002.
[a] Statistically different from control values ($p \leq 0.05$).

5.4.2 Collecting, Receiving, and Storing Samples

It is known that in vivo postmortem metabolism of hydrogen sulfide occurs [11]. It seems reasonable to infer the same for blood and tissue samples. The logistics of collecting and processing samples is therefore of the utmost importance for biomarker analysis. There are many factors involved: the age of samples, how they have been stored, whether they were frozen immediately or after a delay, sample transport routines, and so on.

Testing often must be delayed for practical reasons; for example, even if samples are collected and frozen quickly, the nearest laboratory that can perform the analysis might be quite a distance away [3,8]. Jones has pointed out that it is not unusual to receive postmortem samples some months after the death has occurred [3].

In all, the following factors seem to be key in using these biomarkers:

- Time elapsed between H_2S exposure and taking blood or urine samples must be known.
- Metabolism and biodegradation processes are halted as much as possible by refrigerating (bodies) and freezing (blood and tissue samples).

5.4.3 Stability of Blood Values

H_2S has a short lifespan in the blood. The instability of sulfide ion makes it a less-than-ideal analyte, difficult to detect because the body rapidly oxidizes it to thiosulfate and sulfate [8,14]. Wang has pointed out that H_2S in the blood can be scavenged by methemoglobin to form sulfhemoglobin—also by metallo- or disulfide-containing molecules such as catalase, oxidized glutathione, or horseradish peroxidase [46].

Working with rats exposed for 3 hours to significant levels of H_2S (up to 400 ppm), Dorman et al. [15] found that sulfide and sulfide metabolite disappeared very quickly in lung tissue samples, once the H_2S exposure stopped. Lung sulfide and sulfide metabolite levels increased during the 3-hour H_2S exposure, but within 15 minutes of the exposure stop, they returned to preexposure (endogenous) levels (see Section 5.4.1.2).

5.4.3.1 Decrease in Blood Thiosulfate Values before Death

If death from H_2S intoxication does not occur quickly after exposure, then blood sulfide and blood thiosulfate levels can be expected to decrease over time. In the findings of Yalamanchili and Smith [2], reported in Table 5.2, the initial (antemortem) blood thiosulfate level was 3.5 µg/mL; in a blood sample taken after death, thiosulfate was not detected. In this case, the patient died the morning after H_2S exposure.

5.4.3.2 Increase in Sulfide Values after Death due to Biodegradation

Over 35 years ago, McAnalley et al. [17] remarked that decomposing protein produces hydrogen sulfide and that there is thus a possibility that a decomposed body would give a false positive for hydrogen sulfide. It has been established that blood sulfide concentrations, and certain tissue sulfide concentrations, can increase after death as biodegradation—and microbial hydrogen sulfide production—proceeds [8,11,14,46].

This problem can be seen in the findings of Kage et al. [6], reported in Table 5.2. In that case, a blood sample collected in the hospital 4 hours after death showed 0.958 µg/mL sulfide. Another blood sample, taken at the forensic autopsy 24 hours after the accident, showed 30.334 µg/mL sulfide. The cadaver was kept at 0°C until the autopsy; this is rather worrying, since the general understanding is that refrigeration will prevent biodegradation.

Nagata et al. [47] studied the postmortem changes in sulfide concentrations of body tissues, for rats exposed to 550–650 ppm H_2S and an unexposed control group. They found that sulfide concentrations in the blood, liver, and kidneys of both groups of rats increased with the lapse of time after death. Lung, brain, and muscle tissues did not show a significant change. Postmortem changes in sulfide concentrations in (unexposed) human tissues showed the same trends, except that the human blood tissue showed little or no sulfide increase with lapse of time after death (see Table 5.5).

Maebashi et al. have compiled the blood sulfide and blood thiosulfate data from 16 suicides in Tokyo during the "H_2S suicide craze" of 2007–2009. They found no correlation between time between death and autopsy, and concentrations of sulfide or thiosulfate in the blood. However, the authors make an important point: in the "H_2S suicide craze," H_2S was deliberately generated by mixing bath salts and household cleaning products. Extremely high concentrations are generated—Kobayashi and Fukushima demonstrated that

TABLE 5.5

Changes in Sulfide Concentrations of Body Tissues as Time Increases after Death

Tissue	Rat, Exposed to 550–650 ppm H_2S	Rat, No H_2S Exposure	Human, No H_2S Exposure
Liver	INCR	INCR	INCR
Kidneys	INCR	INCR	INCR
Blood	INCR	INCR	N/C
Lung	N/C	N/C	N/C
Brain	N/C	N/C	N/C
Muscle	N/C	N/C	N/C

Source: Data from Nagata, T. et al., *J. For. Sci.*, 35(3), 706, 1990.
Abbreviations: INCR, sulfide concentrations increase; N/C, no significant change.

9950 ppm can be generated from very small quantities (0.1 mL) of bath salt and cleaner—and the mixing and poisoning is done in a small, confined space. The amount of inhaled H_2S is much greater than in typical industrial accidents; Maebashi et al. suggest that the huge amounts of H_2S taken in have a much stronger influence on its concentration in the blood than postmortem factors. In the normal course of things, the authors say, it is reasonable to expect the H_2S concentration in blood to be affected by postmortem interval, and the temperature, as H_2S is produced by putrefaction of sulfur-containing organic substances [14, 36, 47].

5.4.4 Factors Affecting Thiosulfate in Urine

Thiosulfate in the urine promises to be a valuable tool for investigating, among other things, chronic low-level H_2S exposures. At the present time, however, there are several important areas that need clarification:

- The relation between air concentration of H_2S and urinary thiosulfate levels
- The effect of diet
- The relation between duration of exposure and urinary thiosulfate level
- The time after exposure that it takes for the thiosulfate in the urine to reach its peak

This biomarker has tremendous potential; more information on these and similar points would increase its usefulness in setting threshold levels or for biological monitoring [5, 6, 23, 46, 48].

5.4.4.1 Time between H₂S Exposure and Urinary Thiosulfate Peak

The time after exposure that it takes for the thiosulfate in the urine to reach its peak could use more investigation. "15 hours" is often quoted as the peak in the technical literature, with references to Kangas and Savolainen [21]. Jones [3], however, has pointed out that Kangas and Savolainen reported that it peaks sometime between 5 and 15 hours after exposure. In that experiment, a volunteer was exposed to 18 ppm H_2S for 30 minutes, and urine samples were periodically taken. However, no samples were taken between 5 hours and 15 hours, since this was overnight. The "15-hour" sample was the morning void sample, which had accumulated over 10 hours. At 17 hours, levels had returned to normal.

In animal experiments, Kage et al. [49] exposed rabbits to 100–200 ppm of H_2S. They found that thiosulfate in the urine could be detected 24 hours after exposure.

5.4.4.2 Elapsed Time between H₂S Exposure and Death

In the 16 bath salt suicides studied by Maebashi et al. [14], four cases showed higher thiosulfate concentrations in urine than in blood. This is quite unexpected in such cases. The victims are exposed to extremely high amounts of H₂S—far higher than in most industrial accidents—and it seems reasonable to infer that there would be a very short time between exposure and death.

Maebashi et al. speculate that in these four cases, the amounts of H₂S generated were not so high and that death was therefore not near-instantaneous. It has been determined that two of the four cases had as much, or more, starting material as most of the 16 cases in the report [14]. For the other two cases, no information on amounts of cleaning agent and bath salts is available. However, we will never know what really happened.

5.5 What Can These Biomarkers Tell Us?

When it comes to choosing biomarkers, the dividing line seems to be whether the H₂S exposure was fatal or not.

The elapsed time between H₂S exposure and the collecting of blood or urine samples will affect the values measured. In fatal cases, the elapsed time between exposure and death will also affect the biomarker values. In Figure 5.6, point A represents blood and urine samples taken very soon after a nonfatal case of H₂S exposure. Point B represents samples taken a bit later. Point C might represent samples taken several hours after the H₂S exposure. At A we would expect to see appreciable amounts of sulfide and/or thiosulfate in the blood, but no thiosulfate in the urine. At B we would expect to see that thiosulfate in blood would predominate, though thiosulfate in the urine may start to appear. At C the blood may be cleared of thiosulfate or only contain small amounts, and the urine thiosulfate levels should be elevated.

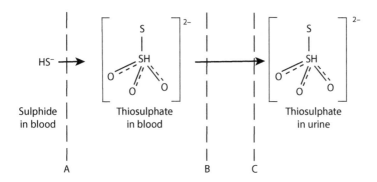

FIGURE 5.6
H₂S moving through the body: samples taken at points A–C after exposure.

5.5.1 Fatal Exposures

There seems to be a general consensus that in the case of fatalities, blood samples are the most promising to yield useful information about likely cause of death. Elevated levels of thiosulfate in the blood can provide corroboration of acute H_2S intoxication, provided that

- The bodies are refrigerated between death and autopsy.
- The specimens are frozen in airtight containers until analysis [5,9,12,14,47,50].

Urine samples might yield useful clues about whether death was instantaneous or occurred after a period of unconsciousness. In cases where the worker was not discovered until some hours after the accident, this may aid in reconstructing what happened.

5.5.2 Nonfatal Exposures

If the H_2S incident does not result in fatalities, then urine thiosulfate is the appropriate biomarker to examine. Blood samples are expected to yield little information, because the metabolism of the survivors will simply clear away sulfide and thiosulfate from the blood too quickly.

Where H_2S exposure is known or suspected to have occurred, Jones [3] recommends multiple samples: as soon as possible after the exposure and additional samples within the 2–15-hour window after the accident. This is not guaranteed to find the peak or maximum amount; but as Jones points out, if you want to determine whether H_2S was a likely causal agent, then it is sufficient to capture the increase in urine thiosulfate levels. The timing of the urine samples, relative to the accident, is important: there is a window after the H_2S exposure, within which thiosulfate will appear in the urine. However, knowing the exact maximum within that window will not add more information or corroborative evidence.

5.5.3 Monitoring or Studies

Urine thiosulfate promises to be a useful tool for certain applications:

- Routine occupational monitoring
- Studies of chronic low-level exposure

By "routine occupation monitoring" we do not mean that this would provide warning of, or protection against, levels that are immediately dangerous to life and health. Occupational monitoring of urine thiosulfate is appropriate only for lower "nuisance" levels, perhaps up to 100 ppm. Jones [3] recommends postshift samples for routine occupational monitoring.

This biomarker also has great potential in studies of chronic low-level H_2S exposure. The public health concerns of chronic low-level exposure is an important question for various groups:

- Communities living in geothermally or volcanically active areas, such as Rotorua, New Zealand, or Reykjavik, Iceland
- Communities with H_2S-generating industries, such as towns with kraft pulp mills
- Workers whose jobs routinely expose them to low levels of H_2S, such as sewage workers

In the following section, we present results from two such studies. We are confident that this biomarker will prove its importance, and fully expect to see more of these in the future.

5.6 Studies Utilizing These Biomarkers

There are numerous studies comparing the health of communities exposed to H_2S to that of unexposed communities. The use of biomarkers, especially thiosulfate in urine, has begun to make an appearance in these studies. In Table 5.6 we give a few examples.

Durand and Weinstein [22] exposed eight volunteers to low doses of H_2S (range: ppb to low ppm). They measured urine thiosulfate before and after the H_2S exposure. The mean thiosulfate concentrations in urine increased from 4.6 to 11.5 µmol/L following exposure (see Table 5.6). Interestingly, the increase was not consistent between individuals. Their findings support the idea that low-level exposures to H_2S can be confirmed by thiosulfate in urine. But it also suggests that the response of this biomarker may vary significantly from one individual to another.

TABLE 5.6

Studies Using Urine Thiosulfate as Biomarker for Low-Level H_2S Exposure

Study	Group	Thiosulfate Measured in Urine	Thiosulfate Normalized against Creatinine
Durand and Weinstein [22]	Before H_2S exposure	0.0046 µmol/mL (0.516 µg/mL)	7.2 µmol/mol
	After H_2S exposure	0.0115 µmol/mL (1.289 µg/mL)	9.8 µmol/mol
Farahat and Kishk [23]	Administrative workers (no exposure)		25.1 µmol/mg creatinine
	Sewer workers (chronic 10 ppm exposure)		50.6 µmol/mg creatinine

Farahat and Kishk [23] performed a study of 33 sewer workers versus a control group of 30 administrative workers in Cairo, Egypt. The sewer workers, whose work included inspecting, repairing, and extracting sediments, were exposed to low levels of H_2S 3–4 times per week; sampling done for the study measured 8.8–10.5 ppm. The group that was exposed to H_2S performed significantly worse in the battery of neurophysiological and neuropsychological tests (memory defects, lack of concentration, lower verbal fluency) and had a significant increase in urinary thiosulfate, compared to the control group. However, they note that the increase in thiosulfate levels—50.6 μmol/mg creatinine for the exposed group and 25.1 μmol/mg creatinine for the control group—did not correlate with any of the applied neurophysiological or neuropsychological tests or the duration of work. The authors suggest that thiosulfate can be useful as an exposure marker, but that it cannot be used as a screening test for cognitive impairment after H_2S exposure.

5.7 Examples of the Accidents in Table 5.2

Table 5.2 lists the biomarkers reported from 24 accidents. To describe each accident in detail would not, we feel, benefit the reader. However, in this section, we describe a few cases, which illustrate different ways H_2S incidents related to water and wastewater can arise and the speed with which the unwary can be overcome.

5.7.1 New Jersey, Construction of Municipal Sewage Pumping Station

Snyder et al. [26] reported on a New Jersey case of multiple H_2S intoxications, involving the construction crew laying the foundation for a municipal sewage pumping station. Thirty-seven people were evaluated for emergency treatment, six were admitted to hospital, and one person was pronounced dead at the hospital—most tragically, one of the rescuers. Serum samples were taken of the nine most-severely exposed people and tested for sulfide and thiosulfate. Sulfide was not detected in any of the samples. They report a relationship between serum thiosulfate levels and adverse health effects, as shown in Table 5.7. They attribute the nondetection of sulfide to two causes:

1. The rapid biotransformation of hydrogen sulfide—sulfide in oxygenated blood has a 15–20-minute lifetime [26,51].
2. The lipophilicity of hydrogen sulfide, which would drive it to concentrate in the erythrocytes (which were not saved).

TABLE 5.7

Sulfide and Thiosulfate Measured in Serum Samples

Worker/Officer	Thiosulfate (μg/mL)	Sulfide	Health Effects
Worker 2	1.3	Not detected	Recovery in 48 hours, after oxygen therapy and supportive care
Worker 1	2.0	Not detected	Prolonged illness due to severe acute neurological and pulmonary injury
Police officer 1	12.0	Not detected	Immediate fatality

Source: Data from Snyder, J.W. et al., *Am. J. Emerg. Med.*, 13(2), 199, 1995.

5.7.2 Spain, Silo for Water Purification Sludge

Nogué et al. [33] have described a Barcelona case involving a sludge silo receiving sludge from water purification stations. In that case, a truck dumped 27 tons of sludge into the silo. H_2S that had accumulated inside the silo was pushed upward and spilled over the top. While the truck was unloading, a worker below the silo was checking a pump. As he was climbing the stairs—adjacent to the silo—to leave the area, he was overcome by the descending H_2S cloud and lost consciousness. Two coworkers went to his aid but also lost consciousness. When firefighters reached the scene, all three men were dead. Blood levels of sulfides of the three victims were 2.48, 14.9, and 18.1 mg/L, respectively (taken 24 hours after the accident).

5.7.3 France, Domestic Poisoning due to Faulty Drains

The domestic accident in Table 5.2 is reported by Sastre et al. [35]. A man came home and found his wife dead in front of the kitchen sink. In a highchair near the sink was the couple's nine-month-old daughter, also dead. In the sink was a plunger, and the first responders reported a strong smell of rotten eggs. An investigation led to the conclusion that the two deaths were due to defective maintenance of the apartment building's pipes and drains and an incorrect assembly of the sink siphon; this led to stagnation of wastewater and formation of a pocket of H_2S. The pocket of H_2S was presumably released when the woman tried to clear the sink pipe by using the plunger. This caused a massive and sudden release of H_2S to rise up out of the sink's drain. The two victims, being right by the sink, were overwhelmed.

Sulfide was not found in the blood of either victim. Possibly it had had time, between exposure and autopsy, to convert to thiosulfate (which was not measured). Tissues samples were also taken of the lungs of both victims; analysis showed high levels of H_2S in the lung tissue, 1.46 mg/kg for the mother and 1.92 mg/kg for the daughter.

References

1. Gabbay, D. S., De Roos, F., and Perrone, J. (2001). Twenty-foot fall averts fatality from massive hydrogen sulfide exposure. *The Journal of Emergency Medicine,* 20(2), 141–144.
2. Yalamanchili, C. and Smith, M. D. (2008). Acute hydrogen sulfide toxicity due to sewer gas exposure. *The American Journal of Emergency Medicine,* 26(4), 518, e5.
3. Jones, K. (2014). Case studies of hydrogen sulphide occupational exposure incidents in the UK. *Toxicology Letters,* 231(3), 374–377.
4. Peng, H., Cheng, Y., Dai, C., King, A. L., Predmore, B. L., Lefer, D. J., and Wang, B. (2011). A fluorescent probe for fast and quantitative detection of hydrogen sulfide in blood. *Angewandte Chemie International Edition,* 50(41), 9672–9675.
5. Milby, T. H. and Baselt, R. C. (1999). Hydrogen sulfide poisoning: Clarification of some controversial issues. *American Journal of Industrial Medicine,* 35, 192–195.
6. Kage, S., Takekawa, K., Kurosaki, K., Imamura, T., and Kudo, K. (1997). The usefulness of thiosulfate as an indicator of hydrogen sulfide poisoning: Three cases. *International Journal of Legal Medicine,* 110(4), 220–222.
7. Singh, A. and Sharma, B. R. (2008). Hydrogen sulphide poisoning: A case report of quadruple fatalities. *Journal of Punjab Academy of Forensic Medicine & Toxicology,* 8(1), 38–40.
8. Knight, L. D. and Presnell, S. E. (2005). Death by sewer gas: Case report of a double fatality and review of the literature. *The American Journal of Forensic Medicine and Pathology,* 26(2), 181–185.
9. Christia-Lotter, A., Bartoli, C., Piercecchi-Marti, M. D., Demory, D., Pelissier-Alicot, A. L., Sanvoisin, A., and Leonetti, G. (2007). Fatal occupational inhalation of hydrogen sulfide. *Forensic Science International,* 169(2), 206–209.
10. Ago, M., Ago, K., and Ogata, M. (2008). Two fatalities by hydrogen sulfide poisoning: Variation of pathological and toxicological findings. *Legal Medicine,* 10(3), 148–152.
11. Policastro, M. A. and Otten, E. J. (2007). Case files of the University of Cincinnati fellowship in medical toxicology: Two patients with acute lethal occupational exposure to hydrogen sulfide. *Journal of Medical Toxicology,* 3(2), 73–81.
12. Ballerino-Regan, D. and Longmire, A. W. (2010). Hydrogen sulfide exposure as a cause of sudden occupational death. *Archives of Pathology & Laboratory Medicine,* 134(8), 1105.
13. Tominaga, M., Ishikawa, T., Michiue, T., Oritani, S., Koide, I., Kuramoto, Y., Ogawa, M., and Maeda, H. (2013). Postmortem analyses of gaseous and volatile substances in pericardial fluid and bone marrow aspirate. *Journal of Analytical Toxicology,* 37(2), 147–151.
14. Maebashi, K., Iwadate, K., Sakai, K., Takatsu, A., Fukui, K., Aoyagi, M., Ochiai, E., and Nagai, T. (2011). Toxicological analysis of 17 autopsy cases of hydrogen sulfide poisoning resulting from the inhalation of intentionally generated hydrogen sulfide gas. *Forensic Science International,* 207(1–3), 91–95.
15. Dorman, D. C., Moulin, F. J. M., McManus, B. E., Mahle, K. C., James, R. A., and Struve, M. F. (2002). Cytochrome oxidase inhibition induced by acute hydrogen sulfide inhalation: Correlation with tissue sulfide concentrations in the rat brain, liver, lung, and nasal epithelium. *Toxicological Sciences,* 65(1), 18–25.

16. Ellenhorn, M. J. and Barceloux, D. G. (1988). Hydrogen Sulfide, in *Medical Toxicology: Diagnosis and Treatment of Human Poisoning*, eds. M. J. Ellenhorn and D. G. Barceloux. Elsevier: New York, pp. 836–840.

17. McAnalley, B. H., Lowry, W. T., Oliver, R. D., and Garriott, J. C. (1979). Determination of inorganic sulfide and cyanide in blood using specific ion electrodes: Application to the investigation of hydrogen sulfide and cyanide poisoning. *Journal of Analytical Toxicology*, 3(3), 111–114.

18. Chaturvedi, A. K., Smith, D. R., and Canfield, D. V. (2001). A fatality caused by accidental production of hydrogen sulfide. *Forensic Science International*, 123(2), 211–214.

19. Kage, S., Nagata, T., and Kudo, K. (1991). Determination of thiosulfate in body fluids by GC and GC/MS. *Journal of Analytical Toxicology*, 15(3), 148–150.

20. Kawanishi, T., Togawa, T., Ishigami, A., Tanabe, S., and Imanari, T. (1984). Determination of thiosulfate in human urine and plasma by high performance liquid chromatography with a dual electrochemical detector. *Bunseki Kagaku*, 33(7), E295–E300.

21. Kangas, J. and Savolainen, H. (1987). Urinary thiosulphate as an indicator of exposure to hydrogen sulphide vapour. *Clinica Chimica Acta*, 164(1), 7–10.

22. Durand, M. and Weinstein, P. (2007). Thiosulfate in human urine following minor exposure to hydrogen sulfide: Implications for forensic analysis of poisoning. *Forensic Toxicology*, 25(2), 92–95.

23. Farahat, S. A. and Kishk, N. A. (2009). Cognitive function changes among Egyptian sewage network workers. *Egyptian Journal of Occupational Medicine*, 33(2), 253–270.

24. Chwatko, G. and Bald, E. (2009). Determination of thiosulfate in human urine by high performance liquid chromatography. *Talanta*, 79(2), 229–234.

25. Ikebuchi, J., Yamamoto, Y., Nishi, K., Okada, K., and Irizawa, Y. (1993). Toxicological findings in a death involving hydrogen sulfide. *Nihon hoigaku zasshi (The Japanese Journal of Legal Medicine)*, 47(5), 406–409. (In Japanese with English summary.)

26. Snyder, J. W., Safir, E. F., Summerville, G. P., and Middleberg, R. A. (1995). Occupational fatality and persistent neurological sequelae after mass exposure to hydrogen sulfide. *The American Journal of Emergency Medicine*, 13(2), 199–203.

27. Ikeda, N., Kage, S., Ito, S., and Kishida, T. (1999). A fatal case of hydrogen sulfide poisoning in a geothermal power plant. *Occupational Health and Industrial Medicine*, 1(40), 18.

28. Kage, S., Kashimura, S., Ikeda, H., Kudo, K., and Ikeda, N. (2002). Fatal and nonfatal poisoning by hydrogen sulfide at an industrial waste site. *Journal of Forensic Sciences*, 47(3), 652–655.

29. Kage, S., Ikeda, H., Ikeda, N. Tsujita, A., and Kudo, K. (2004). Fatal hydrogen sulfide poisoning at a dye works. *Legal Medicine*, 6(3), 182–186.

30. Nikkanen, H. E. and Burns, M. M. (2004). Severe hydrogen sulfide exposure in a working adolescent. *Pediatrics*, 113(4), 927–929.

31. Daldal, H., Beder, B., Serin, S., and Sungurtekin, H. (2010). Hydrogen sulfide toxicity in a thermal spring: A fatal outcome. *Clinical Toxicology*, 48(7), 755–756.

32. Bassindale, T. and Hosking, M. (2011). Deaths in Rotorua's geothermal hot pools: Hydrogen sulphide poisoning. *Forensic Science International*, 207(1), e28–e29.

33. Nogué, S., Pou, R., Fernández, J., and Sanz-Gallén, P. (2011). Fatal hydrogen sulphide poisoning in unconfined spaces. *Occupational Medicine*, 61(3), 212–214.

34. Zuka, M., Chinaka, S., Matsumoto, Y., Takayama, N., Hitomi, Y., Nakamura, H., and Ohshima, T. (2012). Fatal and non-fatal cases of lime sulfide exposure and pathogenetic mechanisms underlying pancreatic injury: Case reports with an animal experiment. *Journal of Forensic and Legal Medicine*, 19(6), 358–362.

35. Sastre, C., Baillif-Couniou, V., Kintz, P., Cirimele, V., Bartoli, C., Christia-Lotter, M.-A., Piercecchi-Mari, M-D., Leonetti, G., and Pelissier-Alicot, A.-L. (2013). Fatal accidental hydrogen sulfide poisoning: A domestic case. *Journal of Forensic Sciences*, 58(s1), S280–S284.

36. Kobayashi, K. and Fukushima, H. (2008). Suicidal poisoning due to hydrogen sulfide produced by mixing a liquid bath essence containing sulfur and a toilet bowl cleaner containing hydrochloric acid. *Chudoku Kenkyukai jun kikanshi (The Japanese Journal of Toxicology)*, 21(2), 183–188. (In Japanese with English summary.)

37. Fujita, Y., Fujino, Y., Onodera, M., Kikuchi, S., Kikkawa, T., Inoue, Y., Niitsu, H., Takahashi, K., and Endo, S. (2011). A fatal case of acute hydrogen sulfide poisoning caused by hydrogen sulfide: Hydroxocobalamin therapy for acute hydrogen sulfide poisoning. *Journal of Analytical Toxicology*, 35(2), 119–123.

38. Iseki, K., Ozawa, A., Seino, K., Goto, K., and Tase, C. (2014). The suicide pandemic of hydrogen sulfide poisoning in Japan. *Asia Pacific Journal of Medical Toxicology*, 3(1), 13–17.

39. Miyazato, T., Ishikawa, T., Michiue, T., Oritani, S., and Maeda, H. (2013). Pathological and toxicological findings in four cases of fatal hydrogen sulfide inhalation. *Forensic Toxicology*, 31(1), 172–179.

40. Truscott, A. (2008). Suicide fad threatens neighbours, rescuers. *Canadian Medical Association Journal*, 179(4), 312–313.

41. Morii, D., Miyagatani, Y., Nakamae, N., Murao, M., and Taniyama, K. (2010). Japanese experience of hydrogen sulfide: The suicide craze in 2008. *Journal of Occupational Medicine and Toxicology*, 5, 28–28.

42. Reedy, S. J. D., Schwartz, M. D., and Morgan, B. W. (2011). Suicide fads: Frequency and characteristics of hydrogen sulfide suicides in the United States. *Western Journal of Emergency Medicine*, 12(3), 300.

43. Chang, S. S., Page, A., and Gunnell, D. (2011). Internet searches for a specific suicide method follow its high-profile media coverage. *Communications*, 168(8), 855–857.

44. Bott, E. and Dodd, M. (2013). Suicide by hydrogen sulfide inhalation. *The American Journal of Forensic Medicine and Pathology*, 34(1), 23–25.

45. Sams, R., Carver, H., Catanese, C., and Gilson, T. (2013). Suicide with hydrogen sulfide. *The American Journal of Forensic Medicine and Pathology*, 34(2), 81–82.

46. Wang, R. (2012). Physiological implications of hydrogen sulfide: A whiff exploration that blossomed. *Physiological Reviews*, 92(2), 791–896.

47. Nagata, T., Kage, S., Kimura, K., Kudo, K., and Noda, M. (1990). Sulfide concentrations in postmortem mammalian tissues. *Journal of Forensic Sciences*, 35(3), 706–712.

48. ATSDR (Agency for Toxic Substances and Disease Registry). (July 2006). Toxicological profile for hydrogen sulfide. U.S. Department of Health and Human Services, Public Health Service, Agency for Toxic Substances and Disease Registry: Atlanta, GA. [online]. Available: http://www.atsdr.cdc.gov/toxprofiles/tp114.pdf. Accessed April 1, 2010.

49. Kage, S., Nagata, T., Kimura, K., Kudo, K., and Imamura, T. (1992). Usefulness of thiosulfate as an indicator of hydrogen sulfide poisoning in forensic toxico-logical examination: A study with animal experiments. *Japan Journal of Forensic Toxicology*, 10(3), 223–227.
50. Kage, S., Kudo, K., and Ikeda, N. (1998). Determination of sulfide, thiosulfate and polysulfides in biological materials for diagnosis of sulfide poisoning. *Japanese Journal of Forensic Toxicology*, 16, 179–189. (In Japanese with English summary.)
51. Beck, J. F., Bradbury, C. M., Connors, A. J., and Donini, J. C. (1981). Nitrite as an antidote for acute hydrogen sulfide intoxication?. *The American Industrial Hygiene Association Journal*, 42(11), 805–809.

6

Methane and Natural Gas

6.1 Introduction

Methane is a highly flammable gas; in fact, it is the primary constituent of natural gas [1]. The greatest hazard posed by methane is explosion or fire. Methane can be ignited by a small spark, such as a non-explosion-proof flashlight.

Methane is colorless and odorless, so there are no simple indicators of its presence until it is too late.

6.1.1 Other Names

Other names for methane, or CH_4, are methyl hydride, marsh gas, swamp gas, and bog gas. Methane is the primary component of natural gas and fire damp.

6.2 Physical and Chemical Properties

Table 6.1 shows some important physical and chemical properties of CH_4.

6.2.1 Explosivity Limits

The lower and upper explosive limits are shown in Table 6.2. As mentioned in Chapter 2, the properties in Table 6.2 do *not* define hard-and-fast boundaries between "safe" and "unsafe" conditions. A methane concentration above the upper explosive limit cannot be taken as "safe," because there is always the possibility of dilution to explosive levels [3,4].

It should also be emphasized that these limits are for two gases only, methane in air. If there are other flammable vapors present, for example, due to unauthorized discharges of solvents to the sewer system, then the limits must be recalculated for the total gas mixture. And adding an inert gas will affect the flammable limits: if CO_2 concentrations are greater than 25% v/v, then methane is rendered inflammable [5]. Carbon dioxide has a

TABLE 6.1

Properties of Methane

CAS number	74-82-8
Relative vapor density (air = 1)	0.6
Color	Colorless
Odor	Odorless
Flammability and explosivity	Extremely flammable (gas/air mixtures are explosive)
Toxicity	Simple asphyxiant
Solubility in water, mL/100 mL at 20°C	3.3

TABLE 6.2

Explosive Range, Methane in Air, 20.9% Oxygen by Volume (%v/v)

Lower explosive limit (LEL)	5.0%
Upper explosive limit (UEL)	15.0%

Sources: Gray, C. et al., *Distribution/Collection Certification Study Guide*, Oklahoma State Department of Environmental Quality, Oklahoma City, OH, 2011; Crowl, D. A., *Chem. Eng. Prog.*, 108(4), 28, 2012.

higher molar heat capacity; it acts as thermal ballast, quenching the flame temperature to a level below which the flame cannot exist [6].

6.2.2 Limiting Oxygen Concentration

The concentration of oxygen in air at which the atmosphere will no longer support combustion is known as the limiting oxygen concentration (LOC) for combustion. LOC values vary, depending mainly on the combustible gas and the inert gas.

The LOC for methane is most commonly reported to be 12% v/v [1,3]. Zlochower and Green [7] have reported that the LOC for methane has been measured at 10.7, 11.1, 11.3, and 12.0 mol.% depending on the size and shape of the test vessel.

6.3 Biological Effects

Fire and explosion are the main dangers associated with methane, but it can also be fatal in other ways.

6.3.1 Asphyxiant

Methane can act as an asphyxiant, displacing oxygen to levels below the minimum needed to sustain life [4,8,9]. Death is due to hypoxia.

The danger is far from hypothetical. Byard and Wilson [10] have reported the case of two children killed in a sewer shaft by methane asphyxia. The two boys, 11 and 12 years old, were found at the bottom of a 37-foot (11.1-meters)-deep sewer shaft. One was declared dead soon after, and the other died in hospital 48 hours later. Analysis of the gas in the shaft revealed a sharp decrease in oxygen with depth:

- 21% O_2 at the surface
- 14.3% O_2 at 5-foot (1.5 meters) depth
- 4.8% O_2 at 10-foot (3 meters) and lower depths

6.3.2 Toxicity

Methane does not produce general systemic effects. There is no evidence that low to moderate exposure levels to CH_4 in air have a toxic effect on humans.

There are some reports of toxic effects at very high levels; the evidence is equivocal, however—and the levels reported would already be fatal due to hypoxia [9].

6.4 Sources of Methane

There are two types of methane in sewers:

1. Thermogenic (geological) methane. This is the methane found in deposits of coal or natural gas.
2. Biogenic methane. This is the methane produced by anaerobic decomposition of organic material.

Biogenic and thermogenic methane can be differentiated by the $\delta^{13}CH_4$ carbon isotope composition [11].

6.4.1 Thermogenic Methane

Thermogenic or geological methane is produced from the burial, compression, and subsequent heating of organic material over geological timescales [4]. The same processes that turn vegetative matter into coal also produce methane, which is why methane is found in coal deposits. Thermogenic methane is the principal component of natural gas.

Thermogenic methane can remain buried in the geological strata until it is disturbed, for example, by tunneling, mining, or tectonic activity. Once it is disturbed, it can migrate up to the surface by following cracks or faults in the rock layers. In some former mining communities, the migrating methane has caused huge problems by seeping into buildings when it reaches the surface. At Arkwright Town in Derbyshire (UK), the methane seeping into houses and other buildings was uncontrollable; in the 1990s the town had to be demolished and rebuilt about a mile away [12].

Thermogenic methane has caused a number of tragedies in the construction of water and sewage tunnels. In Cleveland, Ohio for example, a group of tunnelers digging toward the Kirtland water pumping station hit a natural gas pocket on May 11, 1898. It exploded; all eight tunnelers died from burns. Two months later, on July 11, 1898, a group of 11 men working on the same tunnel hit a natural gas pocket and were killed in the explosion. (The tunnel was sealed off after the second disaster [13].)

6.4.2 Biogenic Methane

Biogenic methane is produced from decomposing organic material in oxygen-poor circumstances, for example, in rising sewer mains, sewage sludge deposits, man-made landfills, rice paddies, bogs, or marshland.

The methane is created by the archaebacterial organisms *methanogenic archaea* (MA). These are often in competition with the sulfate-reducing bacteria (SRB) that produce H_2S. Both MA and SRB are found in the biofilms lining sewers, especially rising mains.

Whether MA or SRB dominate depends on a number of factors, primarily the composition of the wastewater and the ratio of chemical oxygen demand to sulfate (COD/SO_4^{2-}). Below a certain COD/SO_4^{2-} ratio, H_2S production by the SRB will prevail; above that value, methane production by MA will dominate [14,15]. There are various values reported in the literature for the determinant COD/SO_4^{2-} ratio, possibly because the microbial flora is complicated, biofilm structure and activities vary, and metabolic pathways can shift depending on wastewater composition and circumstances [15–18].

6.4.2.1 Variables Controlling Methane Production in Sewers

The environment in a rising sewer main can be ideal for methanogenesis. Conditions that tend to maximize methane production are shown in Table 6.3.

Interestingly, Cavefors and Berndtsson [27] have also found that the pipeline material has an effect on methane production. Pipes made of concrete had significantly more methane production than pipes made of PVC (with similar area/volume ratio). They attribute this to the greater ease with which biofilms can attach to concrete contra PVC.

TABLE 6.3

Conditions That Increase Methane Production in Anaerobic Sewers

Variable	Methanogenesis Increases With
Hydraulic retention time (HRT)	Longer HRT
Chemical oxygen demand (COD)	Higher COD
Biofilm pH	Neutral pH
Temperature	Higher temperature
Biofilm area/liquid volume (A/V) ratio	Higher A/V ratio

Sources: Guisasola, A. et al., *Water Res.*, 43(11), 2874, 2009; House, S.J. and Evison, L.M., *Water Environ. J.*, 11(4), 282, 1997; El-Fadel, M. and Massoud, M., *Environ. Pollut.*, 114(2), 177, 2001; Metje, M. and Frenzel, P., *Environ. Microbiol.*, 9(4), 954, 2007; Guisasola, A. et al., *Water Res.*, 42(6), 1421, 2008; Liu, D. et al., *Biotechnol. Bioeng.*, 100(6), 1108, 2008; Jerman, V. et al., *Biogeosciences*, 6(6), 1127, 2009; Chaosakul, T. et al., *J. Environ. Sci. Health A*, 49(11), 1316, 2014; Sun, J. et al., *J. Environ. Manage.*, 154, 307, 2015.

Gravity sewers may also produce methane, but since oxygen can inhibit methanogenesis, the amounts will be much less compared to rising sewer mains. Methane produced in the deeper anaerobic layers of the biofilm in gravity sewers can be oxidized when it diffuses into the aerobic biofilm layers [15].

The bulk of methane production is expected to take place in pressurized sewers (rising sewer mains) where it may exist as supersaturated in the wastewater [22]. In the lower pressure of the gravity sections, the methane would tend to enter the gas phase. Chaosakul et al. [25] found, however, that transition to the gas phase does not happen immediately; a significant proportion of the methane can still remain dissolved in the liquid phase for a considerable time in gravity sewers.

6.5 What Is the Scope of the Problem?

There is not a great deal reported in the literature on methane measurements in sewers (though what does exist is certainly disturbing). Biogenic methane is generated in sewer systems as a result of decomposition of organic material. Thermogenic methane can end up in sewers when leaks occur in nearby natural gas pipelines.

6.5.1 Biogenic Methane Amounts

Table 6.4 shows the results of two studies in the Netherlands measuring methane emissions from wastewater treatment plants (WWTP). The Kralingseveer WWTP has anaerobic sludge treatment and generates a

TABLE 6.4

Methane Emissions from Wastewater Treatment Plants in the Netherlands

WWTP	CH_4, kg/(PE × Year)	Reference
Papendrecht	0.212	[29]
Kortenoord	0.153	[29]
Kralingseveer	0.297	[28]

Note: PE, population equivalent.

significant amount of CH_4. The other two WWTPs, however, do not have anaerobic sludge treatment [28]. The numbers for these two WWTPs may perhaps give an indication of the scale of biogenic methane that might arise in a sewer system.

From Sweden, Isgren and Mårtensson [30] report wintertime methane formation in small pressurized pipes of approximately 1.2 mg/m/min. Samples taken at manholes in Malmö, Sweden, contained 0.06–0.6 mg/L CH_4. They also note that the concentrations that they saw are much lower than those found in warmer climates.

Some interesting studies in Australia report significant amounts of methane in sewer systems (see Table 6.5).

Pumping stations may be a hot spot for biogenic methane. In a study of 65 pumping stations in Georgia (United States), it was found that methane emissions varied from 1.13 to 11.68 kg CH_4/day [33]. And in a Gold Coast (Australia) study, CH_4 measurements at a pumping station showed concentrations of 1400–2800 ppm [34–36].

TABLE 6.5

Methane Measurements in Australian Sewer Systems

Site	Methane	Notes	Reference
Rising main C27, Gold Coast	Dissolved CH_4: 5–15 mg/L (summer); 3.5–12 mg/L (winter)		[31]
Rising main C016, trial 1	Average CH_4 production rate: 0.7 mg/(L × hour)	HRT = 0–8.7 hours	[32]
Rising main C016, trial 2	Average CH_4 production rate: 3.1 mg/(L × hour)		[33]
Rising main UC09	Average CH_4 production rate: 1.3 mg/(L × hour)	Average HRT = 2.5 hours	[33]
South East Water Limited trial	Up to 5% CH_4 (100% of LEL) sampled in manholes along a sewer receiving mainly industrial wastewaters	Found significant CH_4 generation in gravity sections. Sewer contains food processing wastewater with high COD	[34]

A painter was using an electric sander on a bathtub, in an apartment that had been vacant for over a year. He worked with the door closed to prevent sanding dust from contaminating the rest of the apartment. Suddenly he was engulfed in a ball of orange fire; the fire disappeared as quickly as it had started. Looking down into the tub, he noticed a ring of blue flame around the bathtub drain. He blew on the drain and extinguished the flame. The door of the bathroom had been blown off its hinges by the explosion but fortunately the painter only suffered superficial burns. The incident is believed to have been caused by methane gas originating from the bathtub drain. Since the apartment had been empty for so long, it seems probable that the plumbing's trap mechanism, which normally would prevent the entry of sewer gases, was dry and therefore not functioning [37].

6.5.2 Thermogenic Methane Amounts

In an extensive 2011 survey mapping methane leaks within Boston's city limits, Phillips et al. [38] found six locations where CH_4 concentrations in manholes exceeded 4% at 20°C. The amount contributed from biogenic sources versus natural gas leaking into the sewers was not reported. Overall in their report, Phillips et al. [38] found that the $\delta^{13}CH_4$ signatures indicated fossil fuel from leaking pipelines for a majority of the 3356 leaks identified. In this study, a leak was defined as a spot where each data point exceeded 2.50 ppm (i.e., the 90th percentile of the distribution of all the data). Certain amounts of biogenic methane were found at some sites, but it is not reported if these six sewer locations are among them.

In January–February 2013, Jackson et al. [39] carried out similar surveys measuring methane across Washington, DC. They identified 5893 pipeline leaks. They also sampled the air in the sewers at 19 high-concentration sites and found disturbingly high levels of methane in 12 of the 19: values at these 12 manholes ranged from 4% to 50% methane in the sewer. The $\delta^{13}CH_4$ signatures closely matched the signature of the pipeline gas in the city.

The situation in the United States seems, however, to be improving, due to a drive since 2011 to replace aging pipelines. Lamb et al. [40] report a 25% reduction in pipeline mains leaks, and 16% reduction in service line leaks, among gas distribution companies participating in a 2013 study. They attribute the leak reductions to better pipeline materials, improved sealing of cast iron joints, and enhanced leak detection and repair methods. They note a certain bias in their data: gas distribution companies volunteered to participate in the study. Companies that have recently invested in upgrading their pipelines might be more inclined to take part in a study that quantifies the savings [40].

References

1. Malloy, K. P., Medley, G. H., and Stone, R. (2007). Air drilling in the presence of hydrocarbons: A time for pause. In: *Conference Proceedings, IADC/SPE Managed Pressure Drilling & Underbalanced Operations*, Galveston, TX March 28–29, 2007.

2. Gray, C., Vaughn, J. L., and Sanger, K. (2011). *Distribution/Collection Certification Study Guide*. Oklahoma State Department of Environmental Quality: Oklahoma City, OK.

3. Crowl, D. A. (2012). Minimize the risks of flammable materials. *Chemical Engineering Progress*, 108(4), 28–33.

4. Appleton, J. D. (2011). User guide for the BGS methane and carbon dioxide from natural sources and coal mining dataset for Great Britain. BGS Open Report, OR/11/054. British Geological Survey: Keyworth, UK.

5. NHBC. (2007). Guidance on evaluation of development proposals on sites where methane and carbon dioxide are present. Report No. 04, March 2007. The National House-Building Council (NHBC): Amersham, UK.

6. Drysdale, D. (1985). *An Introduction to Fire Dynamics*, 1st ed. John Wiley & Sons: New York.

7. Zlochower, I. A. and Green, G. M. (2009). The limiting oxygen concentration and flammability limits of gases and gas mixtures. *Journal of Loss Prevention in the Process Industries*, 22(4), 499–505.

8. Knight, L. D. and Presnell, S. E. (2005). Death by sewer gas: Case report of a double fatality and review of the literature. *The American Journal of Forensic Medicine and Pathology*, 26(2), 181–185.

9. Duncan, I. J. (2015). Does methane pose significant health and public safety hazards?—A review. *Environmental Geosciences*, 22(3), 85–96.

10. Byard, R. W. and Wilson, G. W. (1992). Death scene gas analysis in suspected methane asphyxia. *The American Journal of Forensic Medicine and Pathology*, 13(1), 69–71.

11. Schoell, M. (1980). The hydrogen and carbon isotopic composition of methane from natural gases of various origins. *Geochimica et Cosmochimica Acta*, 44(5), 649–661.

12. British Geological Survey. Methane and carbon dioxide from natural sources and mining. Available at: bgs.ac.uk/products/geohazards/methane.html. Accessed December 26, 2015.

13. Bellamy, J. S. (1995). *They Died Crawling*. Gray & Company Publishers: Cleveland, OH, pp. 45–58.

14. Choi, E. and Rim, J. M. (1991). Competition and inhibition of sulfate reducers and methane producers in anaerobic treatment. *Water Science & Technology*, 23(7), 1259–1264.

15. Guisasola, A., Sharma, K. R., Keller, J., and Yuan, Z. (2009). Development of a model for assessing methane formation in rising main sewers. *Water Research*, 43(11), 2874–2884.

16. Rinzema, A. and Lettinga, G. (1988). Anaerobic treatment of sulfate-containing waste water, in *Biotreatment Systems*, ed. D. L. Wise. CRC Press: Boca Raton, FL, Vol. III, pp. 65–109.

17. McCartney, D. M. and Oleszkiewicz, J. A. (1993). Competition between methanogens and sulfate reducers: Effect of COD: Sulfate ratio and acclimation. *Water Environment Research*, 65(5), 655–664.

18. Omil, F., Lens, P., Visser, A., Hulshoff Pol, L. W., and Lettinga, G. (1998). Long-term competition between sulfate reducing and methanogenic bacteria in UASB reactors treating volatile fatty acids. *Biotechnology and Bioengineering*, 57(6), 676–685.

19. House, S. J. and Evison, L. M. (1997). Hazards of industrial anaerobic digester effluent discharges to sewer. *Water and Environment Journal*, 11(4), 282–288.

20. El-Fadel, M. and Massoud, M. (2001). Methane emissions from wastewater management. *Environmental Pollution*, 114(2), 177–185.

21. Metje, M. and Frenzel, P. (2007). Methanogenesis and methanogenic pathways in a peat from subarctic permafrost. *Environmental Microbiology*, 9(4), 954–964.

22. Guisasola, A., de Haas, D., Keller, J., and Yuan, Z. (2008). Methane formation in sewer systems. *Water Research*, 42(6), 1421–1430.

23. Liu, D., Zeng, R. J., and Angelidaki, I. (2008). Effects of pH and hydraulic retention time on hydrogen production versus methanogenesis during anaerobic fermentation of organic household solid waste under extreme-thermophilic temperature (70°C). *Biotechnology and Bioengineering*, 100(6), 1108–1114.

24. Jerman, V., Metje, M., Mandić-Mulec, I., and Frenzel, P. (2009). Wetland restoration and methanogenesis: The activity of microbial populations and competition for substrates at different temperatures. *Biogeosciences*, 6(6), 1127–1138.

25. Chaosakul, T., Koottatep, T., and Polprasert, C. (2014). A model for methane production in sewers. *Journal of Environmental Science & Health, Part A*, 49(11), 1316–1321.

26. Sun, J., Hu, S., Sharma, K. R., Bustamante, H., and Yuan, Z. (2015). Impact of reduced water consumption on sulfide and methane production in rising main sewers. *Journal of Environmental Management*, 154, 307–315.

27. Cavefors, J. and Berndtsson, T. (June 2014). Estimation of methane levels in sewer systems. Bachelor Thesis Nr. 14-01. Department of Chemical Engineering, Lund University: Lund, Sweden.

28. Daelman, M. R. J., van Voorthuizen, E. M., van Dongen, L. G. J. M., Volcke, E. I. P., and van Loosdrecht, M. C. M. (2013). Methane and nitrous oxide emissions from municipal wastewater treatment–results from a long-term study. *Water Science and Technology*, 67(10), 235.

29. van Voorthuizen, E., Kampschreur, M., van Loosdrecht, M., and Uijterlinde, C. (2010). Emissies van broeikasgassen van rwzi's. H₂O, 14/15, 30–33.

30. Isgren, M. and Mårtensson, P. (2013). Methane formation in sewer systems. Mater Thesis No. 2013-1. Department of Chemical Engineering, Lund University: Lund, Sweden.

31. Liu, Y., Sharma, K. R., Fluggen, M., O'Halloran, K., Murthy, S., and Yuan, Z. (2015). Online dissolved methane and total dissolved sulfide measurement in sewers. *Water Research*, 68, 109–118.

32. Foley, J., Yuan, Z., and Lant, P. (2009). Dissolved methane in rising main sewer systems: Field measurements and simple model development for estimating greenhouse gas emissions. *Water Science and Technology*, 60(11), 2963–2971.

33. Foley, J., Yuan, Z., Keller, J., Senante, E., Chandran, K., Willis, J., Shah, A., van Loosdrecht, M., and van Voorthuizen, E. (2011). N_2O and CH_4 emission from wastewater collection and treatment systems: Technical report. GWRC Report 2011-30. Global Water Research Coalition: London, UK.
34. Ibrahim, T. (2010). High LELs in a sewer line receiving Industrial wastewaters. Internal Report; South East Water Limited, Gold Coast, Australia (summarized in Foley et al. 2011).
35. Liu, Y., Sharma, K. R., Murthy, S., Johnson, I., Evans, T., and Yuan, Z. (2014). On-line monitoring of methane in sewer air. *Scientific Reports*, 4, 6637.
36. Liu, Y., Ni, B. J., Sharma, K. R., and Yuan, Z. (2015b). Methane emission from sewers. *Science of the Total Environment*, 524, 40–51.
37. Spencer, A. U., Noland, S. S., and Gottlieb, L. J. (2006). Bathtub fire: An extraordinary burn injury. *Journal of Burn Care and Research*, 27(1), 97–98.
38. Phillips, N. G. et al. (2013). Mapping urban pipeline leaks: Methane leaks across Boston. *Environmental Pollution*, 173, 1–4.
39. Jackson, R. B., Down, A., Phillips, N. G., Ackley, R. C., Cook, C. W., Plata, D. L., and Zhao, K. (2014). Natural gas pipeline leaks across Washington, DC. *Environmental Science & Technology*, 48(3), 2051–2058.
40. Lamb, B. K., Edburg, S. L., Ferrara, T. W., Howard, T., Harrison, M. R., Kolb, C. E., Townsend-Small, A., Dyck, W., Possolo, A., and Whetstone, J. R. (2015). Direct measurements show decreasing methane emissions from natural gas local distribution systems in the United States. *Environmental Science & Technology*, 49(8), 5161–5169.

7

Methane Case Study:
The Abbeystead Explosion

7.1 Introduction

On May 23, 1984, a party of 36 visitors, adults and children from St. Michael's on Wyre, and 8 staff members of the North West Water Authority (NWWA) assembled at the Abbeystead valve house for a demonstration. The visitors were assembled to observe water flowing through the valve house; but instead a massive explosion occurred, turning the valve house into a crater. Sixteen people died; no one escaped uninjured.

How could it go so wrong?

7.2 Background

In the late 1960s, the water authorities that would later become the NWWA wanted to increase the amount of water that could be taken out of the River Wyre at the water facilities near Garstang to supply industrial towns further south such as Preston, Wigan, and Blackburn. Building a new reservoir should be avoided; the area between the Lune and Wyre rivers is one of outstanding natural beauty, and the rivers boasted fine fishing.

The solution that was decided upon was to transfer water from the Lune river to the smaller Wyre river. The next several years were filled with investigations and permit applications. In 1973, Parliament granted authority for the transfer scheme; the NWWA was permitted to abstract from the Lune up to 280 megaliters/day under certain conditions.

In 1974, the well-known civil engineering firm Binnie & Partners was awarded the contract to design the waterworks system and supervise its construction.

7.2.1 The Transfer Scheme

The Lune–Wyre water transfer scheme was very simple in concept: to take water from the River Lune and pump it over to the River Wyre.

The waterworks system that was designed to move the water is shown in Figure 7.1.

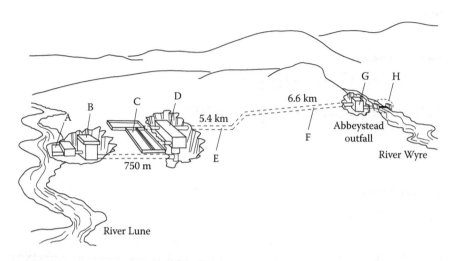

FIGURE 7.1

The Lune–Wyre link. A–D are the Lune intake and pumping station. G–H are the Abbeystead outfall station. A: Forebay. B: Screen house. C: Settling tanks. D: Lune pumping station. E: Quernmore pipeline. F: Wyresdale tunnel. G: Valve house. H: Outfall chambers. (Modified from HSE, The Abbeystead Explosion: A report of the investigation by the Health and Safety Executive into the explosion on 23 May 1984 at the Valve House of the Lune/Wyre Transfer Scheme at Abbeystead, Report by Health and Safety Executive, Her Majesty's Stationery Office, London, UK, 1985.)

The design involved a screening stage, to remove debris, fish, etc. (A and B in Figure 7.1). After the screening, water flowed by gravity to the Lune pumping station (D). At the pumping station the water was moved by low-lift pumps up into the settling tanks (C) and then flowed over a weir into an adjacent balancing tank (whose purpose was to provide a constant head for pumping). A set of high-lift pumps in the pumping station then moved the water south 5.4 kilometers through the steel Quernmore pipeline (E) and then a further 6.6 kilometers through the Wyresdale tunnel (F). In the process the high-lift pumps also lifted the water 100 meters above the Lune intake.

At Abbeystead, the water entered the valve house (G). Here it was distributed among four outfall pipes, which carried the water to an outfall chamber (H) on each bank. There, it was released into the Wyre.

The Wyresdale tunnel and the Abbeystead valve house incorporated a combination of features that were not found at any other water installation in Great Britain. Parameters that would play an important role in the explosion were as follows [1]:

- All the contents of the tunnel—liquid and gas—were emptied into a room with limited ventilation.
- The Wyresdale tunnel was not designed to be watertight.
- Groundwater from the surrounding strata leaked into the tunnel, rather than out of it.

7.2.1.1 The Wyresdale Tunnel

The Wyresdale tunnel has a finished inner diameter of 2.6 meters. It is lined with concrete; at the ends, it has a steel liner (approx. 600 meters at the Abbeystead end and 400 meters at the other end) to strengthen it where the covering ground becomes shallower. The liner was not necessary for the 1984 operating conditions. Instead, it was designed to allow greater water pressures, if the Lune–Wyre transfer scheme was expanded at a later date.

The tunnel is not watertight; the concrete is porous and the tunnel is jointed. Seepage of water or gas could go in either direction, inward or outward, and would depend upon the relative pressures inside and outside the tunnel. Water seepage was estimated to be approx. 0.7–1 million-liters per day* (8 L/s when the tunnel is full, and 12 L/s when empty) [2]. Because the water pressure in the surrounding rocks was much higher than in the tunnel itself, even when the tunnel was full, it was expected that up to 1 million L/day of groundwater would seep into the tunnel from the surrounding strata. This was not expected to pose any problems but would simply increase the water being sent to the Wyre.

At the Abbeystead end, the tunnel divides before reaching the valve house. The western branch is intended to provide access to the tunnel; it goes up to the surface and ends in a blank flange inside an access chamber.

The eastern, water-carrying branch of the tunnel continues under the valve house, over to the Wyre, and goes under that river. On the far side of the river, the tunnel is blanked off to form a dead end (see Figure 7.2).

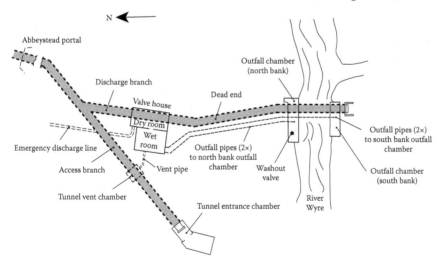

FIGURE 7.2
Tunnel branching at the Abbeystead end. (Modified from HSE, The Abbeystead Explosion: A report of the investigation by the Health and Safety Executive into the explosion on 23 May 1984 at the Valve House of the Lune/Wyre Transfer Scheme at Abbeystead, Report by Health and Safety Executive, Her Majesty's Stationery Office, London, UK, 1985.)

---

* 1 million liters per day is small as a percentage of the water capacity of the tunnel.

7.2.1.2 The Dead End

This cul-de-sac, or dead end, was a preparation for future water needs; it was anticipated that the tunnel would be continued to provide for the growing population on the far side of the Wyre [1,2].

To keep the water in the dead end from stagnating, there were two provisions (not shown in Figure 7.2):

1. A 100-mm dia. circulation pipe back to the weirs in the valve house.

2. A 200-mm dia. washout pipe, running from the discharge pipe dead end to an outfall chamber on the near bank of the river. (The washout valve at the terminus of this pipe is shown in Figure 7.2.)

7.2.1.3 Ventilation at the Abbeystead End

Before the end of the access branch there was a subterranean vent chamber; the tunnel vented into this. There were no access points between Rowton and Abbeystead, so the vent chamber provided venting for the entire length of the tunnel.

The vent chamber was a concrete-lined room, 4.4 meters × 3.6 meters × 2.15 meters high, completely underground and with no ventilation to the outdoors. Within the vent chamber, eight air valves above the pipeline allowed air to escape when the tunnel was being filled or allowed air into the tunnel when it was being emptied. The vent chamber had no connection to the open air; instead, an 800-mm dia. pipe, permanently open, led from the vent chamber into the valve house. In the valve house, the vent pipe discharged just above the grid floor level.

The pipes that brought water into the valve house were also supplied with air valves. These air valves discharged into the wet room, just below the grated ceiling.

All tunnel venting thus led into the valve house, as shown in Figure 7.3.

7.2.1.4 The Valve House and Outfall Chambers

The valve house was divided, at least on paper, into a wet room and a dry room. The dry room, which was directly above the discharge branch of the tunnel, contained the controls for the valves on the water inlets (the "lobster-back" pipe bends), various instruments, and a bathroom. The floor was concrete, except for a section of steel grating over the valve pits. An internal wall with double doors separated the dry room from the wet room. The wet room was the area over the discharge and distribution chambers; it was separated from the water chambers by a steel grating floor.

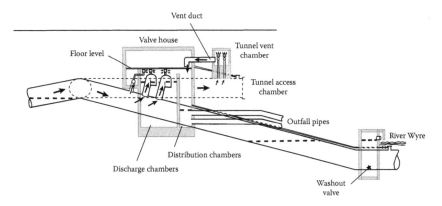

FIGURE 7.3
Venting and gas flows. (Modified from HSE, The Abbeystead Explosion: A report of the investigation by the Health and Safety Executive into the explosion on 23 May 1984 at the Valve House of the Lune/Wyre Transfer Scheme at Abbeystead, Report by Health and Safety Executive, Her Majesty's Stationery Office, London, UK, 1985.)

The wet room was ventilated by a single louvered ventilation panel, 1410 mm × 880 mm, set into the outside wall. The dry room had a 600 mm × 330 mm louvered opening set near the top of the outside wall. In addition, a small electric fan extracted air through metal trunking from below the floor at the rear of the room, where the instruments were situated, and discharged through a separate louvered opening identical to the inlet. In the dry room there were also two small tubular electric heaters that were normally kept switched on [1].

Beneath the dry room ran the discharge branch of the tunnel. From its crown, water flowed through two 1000-mm dia. "lobster-back" pipe bends. The water emptied into two discharge chambers in the wet room. When the discharge chambers were filled, the water flowed over a weir into the four distribution chambers (see Figures 7.4 and 7.5).

Each distribution chamber has an outfall pipe that leads the water to outfall chambers on both sides of the river, two pipes going to the chamber on the near bank and the other two pipes going under the river to the chamber on the far bank. Each outfall chamber has 26 ports to discharge water into the Wyre. This outfall design was used to avoid disturbing the ecology, including fish, on the Wyre.

The discharge ports from the outfall chambers on the river banks are permanently open. The outfall chamber on the near bank, as noted earlier, also received water from the washout valve on the dead end, when the washout valve was opened. This water from the washout valve fed five extra discharge ports, the westernmost (downstream) ports on the near bank.

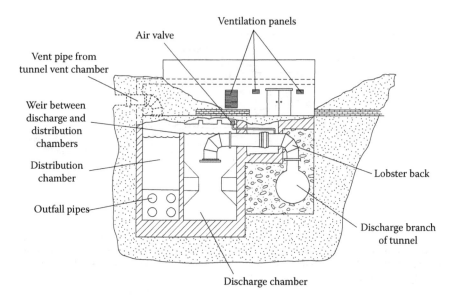

FIGURE 7.4
Elevated view of valve house. (Modified from HSE, The Abbeystead Explosion: A report of the investigation by the Health and Safety Executive into the explosion on 23 May 1984 at the Valve House of the Lune/Wyre Transfer Scheme at Abbeystead, Report by Health and Safety Executive, Her Majesty's Stationery Office, London, UK, 1985.)

7.3 May 1984

The village of St. Michael's on Wyre, with 500 inhabitants, had suffered flooding in recent years, and the inhabitants wondered if the Lune–Wyre transfer scheme could be aggravating the flooding. The NWWA arranged a public demonstration of the operation of the waterworks' valve house. The local parish council was invited to visit the underground facility.

7.3.1 Early and Mid-May

The region was experiencing an exceptionally dry period. For the 17-day period ending in May 23, no water had been piped from the Lune to the Wyre.

7.3.2 May 23, 1984

A group of 36 villagers, including children, and 8 NWWA employees met at the Abbeystead valve house in the evening of May 23.

At 19:12 p.m., a pump was switched on at the Lune pumping station at the other end of the pipeline. Ten minutes later, there was still no water flowing into the valve house. The NWWA's district manager, Alan Lacey, phoned

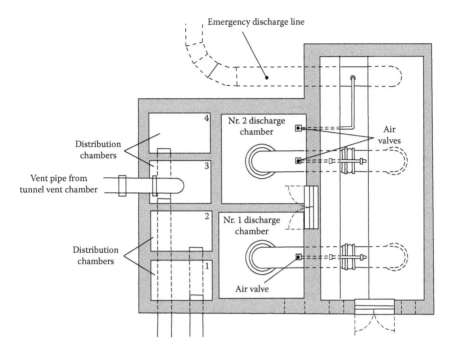

FIGURE 7.5
Plan view of valve house lower level. (Modified from HSE, The Abbeystead Explosion: A report of the investigation by the Health and Safety Executive into the explosion on 23 May 1984 at the Valve House of the Lune/Wyre Transfer Scheme at Abbeystead, Report by Health and Safety Executive, Her Majesty's Stationery Office, London, UK, 1985.)

the pump station to ask for a second pump to be turned on. A few minutes later, a massive explosion occurred in the valve house. The explosion ripped the roof from the pumping plant. Some people were thrown into the water chambers below, as chunks of concrete fell back down and smashed through the steel grating floor. Everyone present suffered blast injuries, crush injuries, and/or burns [1,3,4].

Eight people died on the scene, from severe multiple trauma. Eight more would die in hospital over the next three months, from severe multiple trauma, myocardial infarction, and/or burns sepsis. Of the 28 survivors, 24 had burns covering 5%–65% total body surface area [4].

7.4 The Health and Safety Executive Investigation

The Health and Safety Commission directed the Health and Safety Executive (HSE) to investigate and make a special report in accordance with Section 14(2)(a) of UK Health and Safety at Work Act, 1974. At the same time a joint team from Binnie & Partners/NWWA conducted an independent investigation, which collaborated to a high degree with the HSE investigation [5].

The HSE initially considered several possible causes for the explosion, including

- Naturally occurring methane
- Possible storage of flammable solvents, or liquid petroleum products, in the valve house
- Commercial gas or other flammable liquids leaking from pipelines in the area
- Electrolytic generation of hydrogen in the tunnel
- Terrorist activity

Most of these could be eliminated; what remained was the strong probability that the explosion was caused by gas, most likely methane, which had accumulated in a void in the tunnel and then been pushed into the valve house ahead of the pumped water when the pumping started on May 23. The bulk of the HSE investigation then focused on the following [1]:

1. How, when, and to what extent had a void arisen in the tunnel?
2. What were the nature and source or sources of any gases in the tunnel?
3. How had the gas been transferred to the valve house?
4. How and where had the gas been ignited?

7.4.1 The Void in the Tunnel

The Lune–Wyre link was designed so that it would be full of water when the pumps were running and remain full after the pumps were shut off. The tunnel would be emptied periodically for maintenance, of course; but if operated as intended, no significant void could exist during normal operation. Because ingress of water would continuously occur, on the scale of 1 million liters per day, there should be water flowing through the lobster-backs into the discharge chambers, over the weirs into the distribution chambers, and through the outfall pipes, even when no pumping was going on.

As HSE state in their report on the Abbeystead explosion:

> If the integrity of the system was maintained and none of the washout valves along the tunnel line was open, water should never have been below weir level in the Valve House, and providing the inflow of ground water into the tunnel was at its normal level, there should have been a small continuous outflow of water over the weirs in the Valve House and through all the 52 normal discharge ports on both banks of the river. [1]

7.4.1.1 Indications of a Void

There were a number of indications that the tunnel was not filled with water:

- Survivors of the explosion distinctly remembered, when questioned by HSE afterward, that in the valve house there was no water flowing over the weirs from the discharge chambers to the distribution chambers. In fact, the water level in the discharge chambers was approximately one meter below the weirs before the explosion.
- Survivors who had been down to the river bank just before the explosion said that the only water into the river came out of the five furthest-downstream discharge ports (i.e., those fed by the washout valve).
- A fisherman fishing on this part of the Wyre at the time of the explosion also said that the only water flowing into the river came out of the five discharge ports that were furthest downstream.

The low water level in the valve house distribution chambers, and the lack of flow over the weirs, indicated that the water level in the system had fallen so much that no water was flowing through the lobster-back discharge pipes.

7.4.2 Gases in the Tunnel

When the tunnel was drained from the Rowton end after the explosion, samples of the tunnel water were taken. Analysis showed that the water samples contained dissolved methane, 5–9 mg/L [1].

To investigate the gases in the tunnel, the HSE was aided by HM Inspectors of Mines and Quarries and the Mines Rescue Service of the National Coal Board. Both organizations have experience of working in methane-rich atmospheres.

When the tunnel was (carefully) entered, methane was detected at levels of up to 2%. This is below the 5% lower explosive limit but far above the maximum at which work would be permitted in coal mines. The investigators withdrew until the tunnel had been ventilated by personnel and equipment from the National Coal Board. When it was possible to reenter the tunnel and inspect it, no defects in the lining were found.

Samples of water from the surrounding rock strata entering the tunnel were taken at several points. Methane concentrations varied from a trace, up to 40 mg/L.

7.4.2.1 HSE's Simulation

To test the proposed mechanism of "open washout valve–void in upper tunnel–methane entry from strata," the HSE team together with the Binnie & Partners/NWWA team jointly ran a simulation of the operations leading up to May 23. They refilled the tunnel with water, stopped the pumps, and partially reopened the washout valve. The water level in the tunnel began to fall, and after 12 days a considerable void had formed. The air between the water level and the crown of the tunnel contained 9% methane.

7.4.3 The Sources of Methane

Methane is an odorless, colorless gas; it is both highly flammable and highly explosive. It is produced by the breakdown of organic matter in anaerobic circumstances.

Broadly speaking, methane is of either biogenic or geological origin. Biogenic methane is produced continuously, by natural and anthropogenic systems such as wetlands, rice paddies, organic-rich lake mud, landfills, and wastewater treatment plants [6]. Geological (a/k/a thermogenic) methane is created during the conversion of buried organic matter into coal. It is ancient and found in the coal-bearing geological strata. The gas is held by the coal in an adsorbed state. It is released from coal by mining or by geological disturbances such as faults [7].

Geological methane is subdivided into two types: stress relief and reservoir gas. Stress relief methane is released from carbonaceous rocks when pockets of gas are disturbed by excavation. It is transient; because of its limited volume, it dissipates after its release. Reservoir gas, on the other hand, is found when a layer of methane-rich rock lies beneath a layer of rock that is impervious to methane. The upper layer traps the gas in a reservoir, which may be vast in size. The reservoir caps often leak, for example as a result of tectonic activity. Because the reservoir volume is large and the leak relatively small, the leak can go on over many years.

The three types of methane are shown in Table 7.1.

In order to establish whether the methane was biogenic or geological in origin, samples of the ingress water were analyzed for methane/ethane ratios and isotope identification [5]. The carbon stable isotope ratios (13C/12C) of different methane gases vary according to their formation process, as does the hydrogen stable isotope ratio (2H/1H) [8]. The Isotope Measurements Laboratory at Harwell found that most of the methane was ancient, more than 20,000 years old [3]. This established the origin as geological.

TABLE 7.1

Types of Methane

Origin	Age	Comparative Volume
Biogenic	New	Depends on surface area of wetland, lake, etc.
Geological—stress relief	Very old	Small
Geological—reservoir	Very old	Large to vast

7.4.4 How the Gas Was Transferred to the Valve House

On May 23, 1984, when pumping resumed, the air in the void was pushed along the tunnel and expelled into the valve house.

The HSE confirmed the transfer mechanism in their simulation. After 35 minutes of pumping, they measured 7% methane near the roof of the vent chamber. The concentration in the valve house at that point had risen to 4.8%.

7.4.5 The Source of Ignition

The source of ignition was never determined. There were several possible sources of ignition, because the valve house was not designed as an explosive zone. Equipment, wiring, and lighting were not explosion proof. The upper level, which people occupied, was separated from the lower level by open steel grating.

The minimum ignition energy (MIE) is the minimum energy of an ignition source that is required to ignite the vapor. For methane the MIE is 0.28 mJ, which is a small amount of energy. A spark that is just detectable to the touch contains about 20 mJ of energy [9]. Once the methane in the air reached 5%, almost anything could have sparked an explosion: plugging in a lamp, switching on an electric heater, turning on the fan, a static spark—even, as Orr et al. [5] pointed out, the wrong shoes on the steel grating might have done it.

Could the methane have ignited in the tunnel vent chamber and then spread to the valve house? The HSE investigation found clear signs of a similar explosion in the tunnel vent chamber [1]. Whittingham reports: "Just before the flow of water arrived at the chamber, a blue glow was observed coming from below the metal grating and the visitors felt an intense build-up of heat inside the room. This was followed a few moments later by a huge explosion ..." [10].

We will never know the ignition source, and at the end of the day, it doesn't matter. With a methane-rich atmosphere in a nonexplosive zone, ignition sources were plentiful. Even taking off clothing made of the wrong fibers could have sparked the explosion.

7.4.6 HSE Recommendations

The HSE made several recommendations for tunneled, raw water transfer systems that are not watertight [1]:

1. Systems should be designed so that any gas or water discharged is vented to a safe place in the open air.
2. Where the first recommendation is impracticable, comprehensive testing should be made to determine the nature of any contaminants.
3. Controls for washout valves should be either locked in the closed position or designed such that they cannot be operated by unauthorized persons.

4. During tunneling work, testing should be done for flammable gases.
5. Operators of existing systems should review the possibility of methane being present and take appropriate actions.
6. Operational instructions must include safe systems of work, and these systems should be monitored by management.
7. Training of staff should make them aware of any special features of the water transfer system, including potential hazards and appropriate precautions.
8. The fact that methane is soluble in water and increasingly so at elevated pressure, and that it can be given off by groundwater entering workings, should be widely publicized throughout the civil engineering community and incorporated in professional training.

7.5 The Open Washout Valve

The washout valve existed because the tunnel dead end accumulated silt. The designers' intention was that the dead end would be flushed out periodically by opening the washout valve in the outfall chamber on the near bank of the Wyre. This rather modest eight-inch washout valve would play a significant role in the disaster on May 23, 1984, so it must be discussed at some length.

7.5.1 Changes in Practice

The Manual of Operating Instructions provided by Binnie & Partners stated that the washout valve should be opened periodically to flush out stagnant water from the dead end; how often was not specified. The NWWA began by performing this once a month.

However, it was found that the dead end under the river had a large amount of silting, which discolored the river when the valve was opened once a month. Fishermen complained about deterioration of the waters. In early 1980, the once-a-month flushing was changed to a practice of cracking the valve slightly open whenever pumping was in progress and to close the valve when pumping stopped.

This new practice of cracking open the eight-inch washout valve when the pumps were on, and then shutting it, lasted for a few months. Then the pumping through the Lune–Wyre transfer system came under the remote control of the Franklaw Treatment Works. Abbeystead was not permanently manned [3]. The method was changed again; the washout valve was left permanently cracked open, whether the pumps were running or still [1].

These changes seem to have been made at a very local level; no documentation authorizing or even describing them before the explosion has been reported.

The designers of the system were not consulted about their possible effects. The information detailed above was gathered by the HSE after the explosion, through questioning the surviving NWWA employees who operated the water transfer scheme. As the HSE report states:

> It seems probable that the changes were introduced by NWWA operating staff who saw no reason for referring them to a higher level in the organisation for approval. If this had been done the possibility of partial drainage of the tunnel might have been recognized, and it may well be that the procedures would have been modified to prevent any potential loss of water. No evidence has come to light to suggest that the presence of flammable gas in the tunnel had been envisaged by anyone concerned with the operation of the scheme. [1]

7.5.2 The Result

The HSE estimated that a void in the tunnel of 1425 m^3 was able to form because the washout valve was left open.

An empty space in the tunnel would have lower pressure than the surrounding rock strata. If there was methane in the surrounding rock strata—either methane gas or methane dissolved in the strata water—then the void in the tunnel would

- Cause methane contained in the strata water that had seeped into the tunnel to become gaseous at the lower pressure
- Drive an influx of gaseous methane in the rock strata into the lower pressure of the tunnel

7.5.3 Disproportionate Results

Was the open washout valve the root cause of the explosion? No. It was undeniably a direct cause of the particular circumstances of May 23, 1984. But there was a root cause behind the explosion; and that root cause would have continued to operate until, probably, another set of circumstances arose, which could lead to tragedy.

7.6 The Root Cause of the Accident

The root cause was the ventilation design. Hindsight is always 20/20; but it was a mistake to have a long tunnel whose only ventilation was into an underground chamber with limited contact with open air.

With methane in the strata water, the methane would quickly enter the gaseous phase as the water flowed over the weirs in the valve house. In other words, even if the tunnel were completely filled with water at all times and the operation of the transfer scheme was absolutely flawless, the design still was such that methane would accumulate in the valve house [5,11].

Even if methane had not been present, this was a flawed concept. The design consisted of a long tunnel whose only ventilation was into the subterranean valve house; the underground valve house in turn had only limited open air contact. This effectively turned the valve house into a confined space, apparently without anyone recognizing that fact.

A confined space that is not recognized as such is dangerous; there are any number of hazard scenarios. The air in the valve house can become oxygen deficient if the biological or chemical oxygen demand of water in the valve house or the dead end consumes oxygen. Or at some point in the future, maintenance engineers might decide to apply a coating or lining inside the entire tunnel that would almost certainly introduce significant amounts of flammable solvents during the coating process. The 2007 penstock fire in Colorado, which claimed the lives of five men who were applying a coating, underlines the dangers [12].

There are no details available of how the design error came to be made by the designers or accepted by the client. It appears that it was not discovered by design review processes at either the engineering firm or the client.

7.7 The Engineering Aftermath

The engineering team from Binnie & Partners/NWWA had to consider several technical problems:

- Confirming the source and size of the methane reservoir and the mechanism by which methane entered the tunnel
- Modifications to the water transfer system, to introduce safe and adequate venting of any methane gas entering the tunnel
- Assisting the establishment of safe practices for the future operation of the water transfer system

7.7.1 Identifying the Reservoir and Quantifying Methane Ingress

7.7.1.1 Geological Studies

After further geotechnical and seismic investigations and source rock studies, models were developed to predict potential reservoir formations under the tunnel. The most important was the Pendleside Limestone, a structure more than 1000 meters beneath the tunnel that contained a large area that was believed to be porous—a shallow-water reef type of sediment. This large porous area could hold methane, rather like a huge subterranean sponge. The seismic data identified faulting: large faults slanting down from the surface to form a "V" whose point was in the reef-type sediment. These faults provided gas migration routes from the leaking reservoir up to the tunnel [5].

7.7.1.2 Gas and Water Analyses

In order to ensure an adequate scale to the new venting arrangements, the amount of methane needed to be quantified. A period of intense sampling occurred:

- Gas samples were collected twice weekly at the temporary venting facility, and the volume discharged was recorded daily. The samples were analyzed for methane.
- Water samples were taken daily from the tunnel. These were analyzed for methane.
- Methane gas distribution throughout the tunnel was measured using a sensitive methanometer.
- The distribution of water flows into the tunnel was investigated using a tracer dilution technique.
- Water drainage outflows from the tunnel at Abbeystead and Rowton were sampled.

From all of this data, it was possible to calculate the total methane ingress rate into the empty tunnel each day. It varied over the 26 days of sampling, between 50 and 140 kg/day [5].

The tunnel was refilled, and more samples were gathered, this time to check methane storage in the water.

7.7.1.3 Conclusions about Methane

Significant findings included the following [5,11]:

- Methane can be stored in solution in the water-filled tunnel during periods when water is not being transferred. The zone of high methane concentration—approx. 20 mg/L—in the tunnel water extends in both directions along the tunnel, as the standing period continues.
- Estimated methane storage potential within the tunnel approaches 600 kg dissolved methane if the standing period continues for several months.

- When pumping restarts, about 80% of the methane comes out of solution as the water flows over the weirs in the valve house.
- It is estimated that about 50% of the methane enters the water-filled tunnel as free gas, and the other half is in solution in the ingress water.
- The long-term average rate of methane egress from the tunnel (both dissolved and free gas form) was 8 kg/day.
- The rate of methane entry varies inversely with barometric pressure and with other factors only partially understood.
- The methane is derived from source rocks at considerable depth.
- The methane is migrating upward, probably at a constant rate. The trigger mechanism is unknown; Orr et al. [5] speculate that possibly a minor tectonic event started it.

7.7.2 Criteria for the Redesign

The firm Sir William Halcrow & Partners was appointed by NWWA to advise on the redesign. The known problems facing the engineers were as follows:

1. Methane gas entering the tunnel 2.0–2.5 kilometers from the Abbeystead portal.
2. A large gas reservoir under the area, sufficient to make it likely that methane ingress into the tunnel would continue throughout its service life.
3. Methane, both dissolved and as free gas, would be present in the tunnel near the Abbeystead valve house. Dissolved gas would come out of solution as water cascaded over the weirs in the valve house.
4. The problem of excessive silting in the dead end branch would have to be addressed.

7.7.3 The Modifications Selected

Many options were considered but rejected due to uncertainty of effectiveness, cost, or other practical considerations. The selected modifications included the following:

- Redesigned tunnel venting:
  - A new vent outlet on the main tunnel line, before the access tunnel branch. This would intercept and dispose of most of the free gas in the tunnel before it can reach the valve house.

- Removal of the original tunnel vent chamber and venting system. Instead, new pipework would carry vented gas to the valve house wet room, to be discharged at roof level through open vent stacks.
- Redesigned valve house:
  - Strict separation between wet and dry rooms (e.g., each having their own entrance).
  - The rebuilt wet room was provided with a glass fiber reinforced polymer (GRP) mesh open roof to allow free ventilation.
  - Air valves in the wet room were replaced by open vent stacks discharging at roof level.
  - Security around the perimeter of the valve house and restriction of access was improved by adding a dry moat.
- Zoning of hazardous areas [16,17] was introduced:
  - The valve house wet room is classed Zone 1.
  - The interior of the dewatered tunnel, once steady and safe ventilation conditions are established, is classed Zone 1.
  - The interior of the dewatered tunnel at all other times is classed Zone 0.
  - Restrictions were placed on personnel entering the wet room.
  - The new wet room was designed to be kept free of all electrical wiring and equipment.
- The circulation system for the dead end branch was improved, to lessen the silting problem.
- Operational changes:
  - Excessive buildup of methane dissolved in the tunnel water during standing periods to be avoided by regular tunnel flushing.
  - Continued monitoring of methane ingress by water sampling.
  - Locks were installed on valve chamber covers.
  - Special procedures during dewatering, to provide adequate ventilation.
- Changes to the Manual of Operating Instructions:
  - Information was added explaining why the valve chamber covers must not be opened.
  - An emphasis was added on the importance of adhering to the methods in the manual, and not, for example, cracking open a washout valve.
  - Special emphasis was added on the presence of methane and precautionary measures this entails.

7.8 The Legal Aftermath

The victims, or their representatives, began proceedings (*Eckersley & Others v. Binnie & Others*) against three defendants:

1. Binnie & Partners—The firm of civil engineers responsible for the design of the waterworks system and also for supervising its construction
2. Edmund Nuttall Limited—The contractors who constructed the link, including the tunnel
3. The NWWA—The operators of the system

The legal discussions that followed were protracted. All three parties listed as defendants (the designers, the builders, and the site owners) denied negligence.

The lower court, sitting under Rose J in February and March 1987, found all three liable: designers 55%, constructors 15%, and operators 30% [2]. It was appealed.

In 1988, a majority (Russell LJ and Fox LJ) of the high court found that the designers were 100% liable, on the basis that the evidence entitled the lower court judge to find that there was a risk of methane being present and that this should have been accounted for in the design. This view was not unanimous, however. In a strong dissenting judgment, Bingham LJ held that the evidence presented did not support any finding of negligence against Binnie & Partners.

The high court ordered Binnie & Partners to pay £2.2 million and disallowed further appeal. Settlement was finally made in November 1989 [4,10].

7.8.1 Continuing Duty to Advise?

One fascinating point is that the lower court judge suggested that the designer might be under a continuing duty to advise on new information that might indicate a danger, even after completion of a project and handover to the client. The high court did not really pursue this; possibly because, as Bingham LJ pointed out in his dissenting judgment, this was not the right case:

> What is plain is that if any such duty at all is to be imposed, the nature, scope and limits of such a duty require to be very carefully and cautiously defined. The development of the law on this point, if it ever occurs, will be gradual and analogical. But this is not a suitable case in which to launch or embark on the process of development, because no facts have been found to support a conclusion that ordinarily competent engineers in the position of the first defendants would, by May 1984, have been alerted to any risk which they were reasonably unaware at

the time of handover. There was, in my view, no evidence to support such a conclusion. That being so, I prefer to express no opinion on this potentially important legal question. [2]

This particular principle—a duty to advise even after handover—does not seem to have developed very much in the intervening years [13], but then Bingham LJ did write that it would be gradual, if it occurred.

7.9 Should the Designers and/or Constructors Have Suspected Methane?

Would a reasonable man, careful of the safety of his neighbor, worry about methane after reviewing the information available at the time the tunnel was designed?

It was well known in the 1970s that where you find coal, you also find methane. So would a reasonably competent geologist suspect the presence of coal seams in the rock strata? And should tunneling engineers have worried about methane? To answer, we have to look at the geological information available to the designers.

7.9.1 The Geological Information Available

The designers of the tunnel began by looking for geological information. The British Geological Survey, Great Britain's central repository for geological information, had geological maps for this area—but they were last updated in the late 1800s. Dr. Tony Wadge, leader of the BGS's Upper Paleozoic research group for Lancashire, described them as "a disgrace" [3].

More accurate information would require drilling boreholes. A few were drilled and did not reveal anything alarming. The decision to limit borehole drilling was supported by an independent specialist [14].

7.9.1.1 Modern Geology: A Very Young Science

At this point, it may be useful to examine the state of geology both in the 1880s when the BGS maps were created and in the 1970s when the tunnel was designed. Modern geology is an extremely young science, roughly 50 years or so. In fact, the space age was well underway before geologists had a credible explanation for the Hawaiian volcanoes.

Much of modern geology rests on the understanding of plate tectonics (also called "continental drift") that the earth's crust is divided up into gigantic plates that are in constant motion. Plate tectonics explains earthquakes, volcanoes, mountain chains, and geological strata. The plate tectonics revolution gained important momentum in 1968 when a seminal paper, *Seismology and the New Global Tectonics*, was published by Isacks et al. [15]. Before that, in the 1940s, 1950s, and 1960s—years when the tunnel designers would be engineering students—geologists studied individual rock formations much as early Victorian biologists catalogued plants and animals: they could note the species but not much else. There was no overall theory to make sense of the data or to help predict what lay in the deeper geological layers.

Also, a large amount of the advances over the past 40 years in geological knowledge has been driven by the widespread search for oil, gas, and other fossil fuels. This has introduced awareness of factors that would not have been recognized when the tunnel was designed. In the 1970s, when the tunnel was designed, modern geology was really still in its infancy.

7.9.2 The Oil and Gas Company Borehole

In 1966 Place Oil & Gas Limited sank a borehole approx. three kilometers from the tunnel, in a search for oil to exploit. This is known variously as the Whitmore [1], Whitmoor [11], or Whit Moor [2] borehole. The borehole log was lodged with the Institute of Geological Sciences (now known as the BGS) and classified as "commercial in confidence," meaning that it was not available to anyone else without the permission of Place Oil & Gas.

The borehole results indicated some hydrocarbon presence at 700 feet, 1100 feet, and a larger showing at 3300 feet [2]. This might possibly have alerted the geologist working for Binnie & Partners to at least the need for more exploratory boreholes, if they had known about it. The IGS was not at liberty to divulge the contents of the borehole log or even the name of the company that had drilled it.

After the accident, the UK representative for Place Oil & Gas, Dr. Fothergill, testified that if asked, the company would have provided the tunnel designers with the borehole log information. This was after the explosion, when 16 people had died. It was probably not humanly possible at that point to say anything but yes.

However, drilling deep boreholes is expensive; the data they yield are usually regarded as the intellectual property of the drillers; and certainly Place Oil & Gas did deposit the borehole log with the IGS as "commercial in confidence." Dr. Fothergill also had some strings attached: "If there was an enquiry where the information owned by the oil company was such that there would be a safety element involved in the enquiry, then the oil company would give favourable consideration to releasing such information on a confidential basis" [2].

In other words, the borehole log could have provided the first hint to the tunnel designers that there were safety hazards. If the tunnel designers had made their request *based on those safety hazards*, then they would have been granted access to the borehole log. There is a certain Catch-22 here.

7.9.3 Hints from IGS*

The IGS, which had the results of the Whitmore borehole (but could not divulge them), contacted the NWWA to express concerns about the amount of water ingressing from the rock strata. The IGS was concerned that there might be an inrush of water into the tunnel, which could be disastrous. The water authority passed this on to Binnie & Partners, who met with the IGS to clear up the questions. Minutes were made of this meeting, containing one paragraph that is pertinent to our inquiry:

> The geological sequence consists of two sandstone formations known as the Roeburndale Grit and the Pendle Grit separated by a sequence of silt-stones, sandstones, mudstones and shales which may be much thicker than the 450 ft indicated by the old maps. (The IGS have information from one deep borehole which they cannot divulge.)

The reference to the deep borehole was of course the Whitmore borehole. But were the IGS hinting as hydrocarbons? The designers of the tunnel, reading the minutes, thought that the sentence in parentheses was an explanation of the preceding sentence about the thickness of a certain geological sequence on the old maps. (This surely is the natural reading of the paragraph and in fact was never contested by the IGS.)

After the meeting, Binnie & Partners added an addendum to their tunnel construction specifications. The IGS confirmed in writing that the addendum was a correct account of the geological information available from the IGS.

It seems reasonable to infer that at this point, the IGS were not worried about methane. They took the step of contacting the NWWA to express concern over the amount of water that might enter the tunnel from the rock strata. The overall impression, admittedly 40 years later, is that if the IGS had suspected a safety hazard from a methane reservoir, then they would have conveyed that suspicion to both the water authority and the tunnel designers.

7.9.4 Coal Was Mined Here 200–300 Years Previously

There was little published information about this before 1984. Dr. Tony Wadge of the BGS said in an interview after the accident that local seams of coal had been recorded on maps. Local people had dug pits up to 10 meters deep to extract coal—albeit this was a few centuries ago [3,10]. And the geologist Frank Moseley appears to have made reference to the coal seams in a paper describing the geology of the Lancaster Fells, published in the journal of the *Geological Society* in 1953 [5]. It is not clear how wide a circulation this paper received, especially among civil engineers in the water industry.

Nothing indicates that the engineers at Binnie & Partners, or the NWWA, were familiar with the mining that took place a few hundred years earlier.

---

* The IGS (Institute of Geological Sciences) was renamed the British Geological Survey (BGS) on January 1, 1984.

It should be remembered that they were working in the 1970s, before the age of Internet databases for the scientific literature.

7.9.5 Headaches during Tunneling

After the explosion one of the miners who had worked on the tunnel, Mr. Eddie Kluczynski,* spoke to the press:

> I am sure gas got in ... When I was working in one section of the tunnel ... I got blinding headaches. So did the other men down there with me. We felt sleepy and drowsy. [3]

To be quite fair, these symptoms are not unique to methane, or even symptomatic of methane. When a miner engaged on a tunnel project that used blasting reports these symptoms, carbon monoxide poisoning from the blasting would probably be the first suspect. Carbon monoxide is an infamous by-product of explosive blasting and has caused fatal accidents on more than one sewer construction project. Carbon dioxide would also cause these symptoms and had been known to cause fatalities in mining [7]. Oxygen deficiency from inadequate ventilation could also explain the headaches [1,2].

7.10 Lessons Learned

Why is it important to study Abbeystead in such detail? Whittingham [10] gives a possible answer, in his analysis:

> There is often reluctance on the part of management and industry seriously to question design criteria for equipment, where changes to these criteria would prove expensive or be perceived as impractical. The result is an over-emphasis on the direct causes of the accident, often associated with human error at the operational level, rather than the root causes associated with the design.

7.10.1 Ventilation into Subterranean Valve House

It was a mistake to have a long tunnel whose only ventilation was into an underground chamber with limited contact with open air. The HSE summarized this in one sentence: "If a water discharge system open to atmosphere had been used, an explosion would almost certainly not have occurred" [1].

We do not know how this design came to be proposed, or accepted by the client. Certainly there were specifications for the Lune–Wyre link. But architects and engineers receive information from the client in other ways

---

* Mr. Kluczynski also testified at the trial *Eckersley & Others v. Binnie & Others.*

also—meetings, telephone conversations, and so on—where clients can, consciously or not, let it be known that some criteria in the spec are more important than others. It is possible, for example, that the emphasis on the natural beauty of the area, and the importance of not disturbing the local environment or the fishing, came to mean in everyone's minds that "everything must be underground." But we will never know.

7.10.2 Methane Is Soluble in Water

One fact that emerged from the HSE inquiry and the lawsuit was that water engineers apparently did not know that methane can be soluble in water. At normal temperature and pressure, of course, it isn't very soluble. But if the pressure in the surrounding strata is high, the amount that can be dissolved increases significantly.

There were some references to methane dissolved in water in the published literature, but they did not appear to have had wide circulation. Definitely not among the civil engineering profession that dealt with water supply and water transport systems [1].

7.10.3 Hic Est Dracones*

After the explosion, an exhaustive site investigation into the geology of the strata under the tunnel took place. A methane reservoir was found—but only, as Orr et al. point out [5], because the geologists knew what they were looking for. However, it will not always be possible to undertake such exhaustive site investigations for every tunneling project.

Abbeystead has opened up several new factors for designers of large-scale water transfer projects:

- Methane reservoirs may be at great depth and in unexplored terrain.
- Methane can travel a long way from its source. In gaseous form, it travels more or less upward (though the path will slope to follow a fault). Dissolved in water, there is no such limitation to the direction of movement.

If the geological structure of the strata in which the tunnel will be built is well known, and the risk of methane can be eliminated with confidence, then there is no new problem. But if the underlying stratum is terra incognita, then the situation is less satisfactory. The engineers may have to assume methane presence as a working hypothesis and devise a systematic series of tests during construction to test the methane hypothesis. This will be helpful, though, only if a strategy has already been prepared for the contingency of meeting methane. Instead of "less satisfactory" one is tempted to describe the situation as "quite unsatisfactory."

---

* "Here are dragons"—a term sometimes found on old maps, indicating uncharted seas.

7.10.4 The Value of Inquiry Commissions

The Abbeystead explosion is frequently used as a case study in safe design. An interesting leitmotif running through many of the reports, journal articles, and chapters written about Abbeystead is the wish for a wider use of ad hoc commissions or boards of inquiry to investigate disasters. As Orr et al. [5] put it:

> The situation concerning the advance in knowledge from an engineering accident or disaster is thoroughly unsatisfactory in the UK. A court of law does not unravel the technical facts in a form to advance general understanding. An arbitration, whether settled in or out of court, does not lead to any publication of the conclusions. The expert witnesses are normally prevented from disclosing what they may have learned. There is no forum for reaching an agreed conclusion on the basis of the expert testimony. No one expert is in command of all the material facts, and it is often necessary to undertake appropriate studies before full conclusions can be reached. Abbeystead indicates something of the extent of such studies, and the conclusions yet remain to some degree conjectural.

Do they have a valid complaint? In the court case *Eckersley & Others v. Binnie & Others*, tried by Rose J sitting at Lancaster in February and March 1987, the hearing occupied the court for six weeks; 49 witnesses were called, 11 of them experts; and the judgment ran to some 84 pages of transcript. This is very thorough; and yet—The court can only hear the witnesses selected for it by the interested parties. The witnesses can only answer the questions put to them.

One rather wonders if the truly important questions are always asked, let alone answered.

References

1. HSE. (1985). The Abbeystead explosion: A report of the investigation by the Health and Safety Executive into the explosion on 23 May 1984 at the valve house of the Lune/Wyre Transfer Scheme at Abbeystead. Report by HM Factory Inspectorate, Health and Safety Executive. Her Majesty's Stationery Office: London, UK.
2. Fox, Bingham, and Russell, L.JJ. (1988). Eckersley & Others v. Binnie & Others: Court of Appeal (Civ Div). 18 Con LR 1.
3. Pearce, F. (May 23, 1985). Blowing the roof off Abbeystead. *New Scientist*, 1457, 28–32.
4. Jaffe, W., Lockyer, R., and Howcroft, A. (1997). The Abbeystead explosion disaster. *Annals of Burns and Fire Disasters*, 10(3), 1–4.
5. Orr, W. E., Muir Wood, A. M., Beaver, J. L., Ireland, R. J., and Beagley, D. P. (1991). Abbeystead outfall works: Background to repairs and modifications—And lessons learned. *Water and Environment Journal*, 5(1), 7–18.

6. Guisasola, A., de Haas, D., Keller, J., and Yuan, Z. (2008). Methane formation in sewer systems. *Water Research*, 42(6), 1421–1430.

7. Appleton, J. D. (2011). User guide for the BGS methane and carbon dioxide from natural sources and coal mining dataset for Great Britain. BGS Open Report, OR/11/054. British Geological Survey: Keyworth, UK.

8. Hitchman, S. P., Darling, W. G., and Williams, G. M. (1989). Stable isotope ratios in methane containing gases in the United Kingdom. Technical Report WE/89/30. British Geological Survey: Keyworth, UK.

9. Crowl, D. A. (2012). Minimize the risks of flammable materials. *Chemical Engineering Progress*, 108(4), 28–33.

10. Whittingham, R. B. (2004). *The Blame Machine: Why Human Error Causes Accidents*. Elsevier Butterworth-Heinemann: Oxford, UK.

11. Muir Wood, A. M. (2002). *Tunnelling: Management by Design*. Taylor & Francis: London, UK.

12. CSB. (2010). Xcel energy hydroelectric plant penstock fire. Report No. 2008-01-I-CO, August 2010. U.S. Chemical Safety and Hazard Investigation Board; Washington, DC.

13. Uff, J. (November 2002). Engineering ethics: Principles and cases. *Ingenia*, 14, 53–58.

14. Summerhayes, S. (2010). *Design Risk Management: Contribution to Health and Safety*. John Wiley & Sons: New York.

15. Isacks, B., Oliver, J., and Sykes, L. R. (1968). Seismology and the new global tectonics. *Journal of Geophysical Research*, 73(18), 5855–5899.

16. British Standards Institution. (BSI). (1983). BS5345: Part 2: 1983, British Standard Code of Practice for selection, installation, and maintenance of electrical apparatus for use in potentially explosive atmospheres (other than mining applications or explosive processing and manufacture), London, UK.

17. British Gas (BG). (1986). British Gas Engineering Standard, BGC/PS/SHA1, Code of Practice for Hazardous Area Classifications for Natural Gas. British Gas plc: Leicestershire, UK, p. 2.

8

Case Study, Natural Gas:
The East Ohio Gas Co. Explosion

8.1 Introduction

Friday, October 20, 1944, 19:00 GMT (14:00 EST)

If you were a member of the U.S. 1st Infantry Division or the 30th Infantry Division outside of Aachen, Germany, then you were preparing for the final assault on that city. It would fall the next day.

If you were a member of the combined Filipino and American troops who had landed on Tacloban beach, then you were either sleeping (it was 3 a.m. on Leyte) or waiting for dawn.

But if you were one of 73 employees at the No. 2 Works of the East Ohio Gas Company, or one of many housewives in the Norwood–St. Clair neighborhood of Cleveland, Ohio, then you had less than two hours to live.

8.1.1 Synopsis of the Disaster

At 2:40 p.m. (Cleveland time), a tank containing 1.1 million gallons of liquefied natural gas (LNG) suddenly failed, opening at the side and discharging all of its contents amidst the gas plant and the surrounding residential and industrial neighborhood. The resulting fire spread through the sewer system; before it was over, it had killed 130 people, partially or completely destroyed 15 factories, and consumed 79 houses.

8.2 The Winter Demand for Natural Gas

The East Ohio Gas Company (EOG Co.) was a public utility that distributed natural gas for industrial and domestic customers in Cleveland, Ohio. It had a problem: meeting the demand peaks in the winter. In the summer there was more than enough gas available; but during very cold spells, demand for gas peaked sharply above the normal baseline demand. The utility needed a reliable supply of gas that could be quickly delivered into the distribution system during peaks in demand. The gas storage facilities could not store a sufficient quantity to get through an extremely cold spell. The January 1940 cold wave, for example, caused gas shortages through the eastern United States. In the early 1940s, the EOG Co. had been forced to curtail gas service to customers a number of times [1,2].

The problem was exacerbated by World War II. Cleveland was the fifth largest hub of war-related businesses in the United States [3]. The war industries increased the base-load demand for natural gas.

8.2.1 The Gas Pipelines

The EOG Co. in Cleveland purchased its natural gas from the Hope Natural Gas Co. in West Virginia. It was transported in four 18- and 20-inch (457–510 mm) lines, from a station 150 miles (241 kilometers) away [2,4].

The EOG Co. had made plans for an additional 12-inch (305 mm), high-pressure line to bring gas to Cleveland. However, interesting results were being reported from Cornwell, West Virginia, about a liquefaction process being developed. The plan was accordingly changed.

8.2.2 Liquefied Natural Gas

The solution decided upon was to divert natural gas from the pipeline during low-demand seasons, liquefy it, and store it until needed—liquefied natural gas (LNG).

LNG is natural gas that has been cooled to ca –250°F (–157°C) at near-atmospheric pressure. It is a clear, odorless, noncorrosive liquid that occupies 1/600th its original gas-phase volume.

Today there are approximately 55 local utilities in the United States that own and operate LNG plants [5], but in 1944 there was only one in the world: the EOG Co. No. 2 Works in Cleveland.

8.3 The Liquefaction, Storage, and Regasification Process

The liquefaction, storage, and regasification (LS&R) plant at EOG Co. No. 2 Works was the first commercial facility in the world in which large quantities of LNG were stored at extremely low temperature, –250°F (–157°C), and low pressure, <5 psi (0.34 atm). There had been a pilot plant in Cornwell, West Virginia, run by Hope Natural Gas Co. [4], to develop and test the LS&R process. Much of the information used to design the LS&R plant in

Cleveland resulted from the testing done at the Cornwell plant. Keys findings included the following [6,7]:

- The metals that retained a safe Charpy impact test value at −50°F (−46°C) are pure copper, bronze, Monel metal, red brass, stainless steel, and steel plate with less than 0.09% carbon content and containing nickel over 3½%.
- Cork was found to be the best insulator of the materials trialed.
- Evaporation from the stored liquid was entirely methane. If allowed to stand for long periods, then the liquid left will become increasingly high in ethane and higher hydrocarbons.

During off-peak periods, such as summer, natural gas could be liquefied and stored in liquid form; then during peak demand periods, that is, winter, the liquid gas could be returned to its gaseous state for distribution and use. The LS&R plant allowed large quantities of gas to be stored more economically than the usual gas holder, as 600 ft³ of gas could be reduced to 1 ft³ of liquid.

The LS&R plant at No. 2 Works used a cascade system, where refrigeration is provided by a series of liquids with decreasing boiling points:

1. Ammonia gas is compressed and then cooled with water.
   - The ammonia boiling point is −27°F (−33°C) at atmospheric pressure.
2. The liquefied ammonia is used to condense and refrigerate ethylene.
   - The ethylene boiling point is −152°F (−102°C) at atmospheric pressure.
3. The refrigerated, liquefied ethylene is used to liquefy compressed natural gas at a high pressure.
4. The liquefied natural gas is expanded from a high pressure to a pressure slightly greater than atmospheric. Some natural gas flashes off, providing further cooling.

Each refrigerant has a lower boiling point than the preceding refrigerant, hence "cascade."

8.4 The Physical Plant

The site selected for the LS&R plant was a property in East Cleveland owned by the EOG Co. and appropriately located on the gas distribution system [8]. The No. 2 Works was bounded on the north by railroad lines, on the east by East 63rd Street, on the west by East 61st Street, and on the south by a residential area.

The railroads and manufacturing activities in the area generated a certain amount of ground tremors, but these were not considered significant. The bearing capacity of the ground that would support the tanks was low, but the foundations of the tanks were built to compensate for that. The foundation design remained stable throughout the disaster [8].

In addition to the LS&R plant, the No. 2 Works included workshops and buildings for the company's natural gas business.

8.4.1 The Original LS&R Plant: Three Spherical Tanks

The LS&R plant was started in September 1940 and finished in January 1941. Principal units were a compressor building, cooling tower, heat exchangers, liquefied gas storage tanks, and regasification equipment. The plant was designed to liquefy 4 million cubic feet (STP) of natural gas per day and to regasify 3 million cubic feet (STP) per hour.

The plant originally contained three spherical storage tanks, each with inner diameter of 57 feet and capacity to store a volume of liquid that would convert to 50 million cubic feet of natural gas [4]. A spherical storage tank is shown in Figure 8.1.

The inner shell was made of 3½% nickel steel, and the outer shell was carbon steel. Between the shells was 3 feet (1 meter) of solid cork block insulation on the lower part where the tank rested, and granular cork insulation on the upper part.

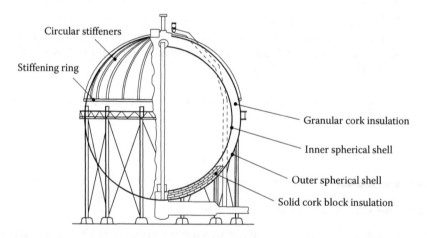

FIGURE 8.1

Spherical storage tank. (Adapted from Elliott, M. A. et al., Report on the investigation of the fire at the liquefaction, storage, and regasification plant of the East Ohio Gas Co., Cleveland Ohio, October 20, 1944, Report of investigations, United States Department of the Interior, Bureau of Mines, Bulletin 3867, Washington, DC, 1946.)

8.4.2 The 4th Tank

After the three spherical tanks were in place and performing satisfactorily, the gas company found in 1942 that even more storage capacity was needed. It was decided to build a fourth tank that could store liquid equating to 100 million cubic feet (2.8 million cubic meter) of natural gas—twice the amount of one of the spherical tanks. There were concerns, however, about the integrity of a spherical tank of such size. The chief engineer of the Pittsburgh-Des Moines Steel Co. (PDM), J.O. Jackson, pointed out that filling and emptying the sphere would cause repeated flexure in the region of the maximum bending stress, which might result in fatigue failure [4].

The fourth tank would have a different design: it would be a vertical cylindrical toro-segmental tank. The bottom resembled "a shallow circular dish with a trough around it" [9], and the walls were two concentric cylinders. The inner cylinder was 42 feet (13 meters) high and 70 feet (21 meters) in diameter. The outer cylinder was 51 feet (16 meters) high and 76 feet (23 meters) in diameter. The fourth tank is shown in Figure 8.2.

Because cork was a war material and difficult to obtain [4,8,9], the insulation was rock wool. The insulation was three feet (one meter) thick, as in the spherical tanks. One million pounds of rock wool was used. The rock wool could not support the weight of the inner shell, as the solid cork supported

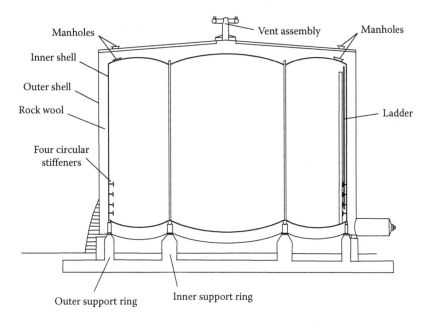

FIGURE 8.2

Cylindrical storage tank. (Adapted from Elliott, M. A. et al., Report on the investigation of the fire at the liquefaction, storage, and regasification plant of the East Ohio Gas Co., Cleveland Ohio, October 20, 1944, Report of investigations, United States Department of the Interior, Bureau of Mines, Bulletin 3867, Washington, DC, 1946.)

the inner shell in the spherical tanks. Instead, the inner shell was supported on two concentric circular support rings of reinforced concrete. The weight of the inner shell and its contents were transmitted to the concrete support rings by 30 wood posts on the inner ring and 60 wood posts on the outer ring. Each wood post was topped with a metal shoe, which fastened to circular girders attached to the inner shell [4,8].

8.4.2.1 The 3½% Nickel Steel

When ordinary steels are exposed to the extremely low temperatures used in LNG, they become embrittled. Only a steel that remains sufficiently ductile to avoid cracking at such a temperature can be used for LNG.

One of the standard tests used for such exposures is the Charpy impact test, which had been issued by the ASTM as a standard test method in 1933. The Charpy impact test measures the energy absorbed when a notched metal sample is struck by a pendulum. The more energy absorbed, the tougher the metal and the less likelihood that it will fail catastrophically if subjected to mechanical shocks or blows. The steel used for the three spherical tanks had a Charpy impact test of 2–4 ft-lbs (2.7–5.4 Nm) [9]; this was not terribly high and was the subject of some discussion. However, the Cornwell results—though produced at the higher temperature of –50°F (–46°C)—seemed to indicate that 3½% nickel steel could be satisfactory. The same steel was used for the fourth tank.

It is worth remembering that fracture mechanics in general was not well understood at this time. The implications of the S-curve—the ductile-to-brittle transition that occurs with temperature decrease in steels—for design were perhaps not fully understood. Certainly if the fracture mechanics and ductile-to-brittle transition had been well known and understood, then the Liberty ships that broke in half during the war would never have been sent into the cold North Atlantic without postweld heat treatment.

8.4.2.2 The Cracked Bottom

After the tank was erected and tested, it cooled prior to being taken into operation. During the cooling a three- to four-foot long crack developed in the bottom [9]; the affected section was replaced and some modifications were done (additional thermocouples, and copper tubing rings to disperse liquid were added). The hydrostatic tests were repeated and the tank passed. The second cooling was successful [2]. The tank was ready for operation in September 1943.

8.4.2.3 The Settling Insulation

Since the No. 2 Works was situated in an area of industrial activity and railroad traffic, the ground sometimes had discernible vibrations. This was not a cause for concern regarding the structure of the tanks (as shown by the

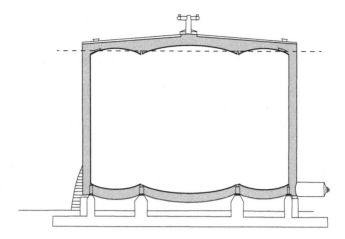

FIGURE 8.3
Shaded region is rock wool insulation. Broken line at approximately four feet (1.2 meters) from the top.

vibration testing performed in the postdisaster investigation) [8]. However, constant small vibrations might have caused the rock wool insulation to continually settle.

Frost spots were noticed on the top and bottom of Tank 4. This had also happened on one of the spherical tanks. The frost spots were caused by settling insulation and disappeared when more insulation was added.

The frost spots that occurred from time to time at the top of Tank 4 were within four feet (1.2 meters) of the top of the tank. Figure 8.3 shows the regions filled with rock wool; the broken line is roughly four feet (1.2 meters) from the top.

The chief engineer at PDM, Mr. J.O. Jackson, advised the EOG Co. to eliminate the frost spot at the bottom of Tank 4 by continuously ventilating it with warm air. It disappeared whenever this was done and reappeared when the ventilating was stopped [8].

8.4.3 Leak Control

The EOG Co. took precautions to control small and moderate rates of LNG spillage and leakage, based on the following assumptions [4,8]:

- Both the tank designers and the EOG Co. assumed that any catastrophic collapse would be preceded by a small leak.
- A small leak would be noticed and corrected before it became serious.
- The idea of a sudden massive spill was not credible. (In their 1978 report on LNG safety [8], the General Accounting Office notes that this assumption is made today in designing most dikes.)

8.4.3.1 Why Was There No Dike or Embankment at First?

Originally there was no dike or embankment around the storage tanks or the LS&R area at Plant No. 2. The codes did not require it. In the National Fire Protection Association's 1943 *National Fire Codes for Flammable Liquids, Gases, Chemicals and Explosives*, section "Regulations for the design, installation, and construction of containers and pertinent equipment for the storage and handling of liquefied petroleum gas," the regulation governing dikes was as follows:

> Because of the pronounced volatility of liquefied petroleum gases, dikes are not normally effective, hence their general requirement is not justified as in the case of gasoline or similar flammable liquids. When, however, in the opinion of the inspection department having jurisdiction, owing to the slope of the ground or other local conditions, above-ground containers are liable, in case of rupture or overflow, to endanger adjacent property, each container shall be surrounded by a dike of such capacity as may be considered necessary to meet the needs of the situation under consideration by the aforesaid inspection department, but in no case more than the capacity of the container in question." [10] (Note: In their report of the Bureau of Mines investigation into the East Ohio Gas Co. fire, the authors note that the phrase "...in no case more than the capacity..." was believed to be a misprint, and should have been "less" instead of "more." [4])

8.4.3.2 Dams Added

As mentioned earlier, the original plan for the LS&R plant did not include dikes or embankments around the storage tanks. It was later decided, however, to add concrete dams to the spheres and Tank 4, in case of minor leaks.

For the spheres the concrete dam was 58 feet (18 meters) in diameter and 4 feet 9 inches (1.4 meters) high and skirted. For Tank 4 it was 85 feet (26 meters) in diameter and 7 feet (2.1 meters) high and skirted [2,4].

In addition to the dikes, each tank had an individual drain leading to an overflow tank [4,8].

8.5 The Events of Friday, October 20, 1944

The liquefaction operation of the LS&R plant was beginning to be shut down for the winter; the tanks, including Tank 4, were filled to capacity. Shutting down the liquefaction operation was a normal procedure, consisting of withdrawing from the system the ethylene and ammonia, shutting down the engines, and stopping the flow of gas. It began at about 2 p.m. [2,4,8,9].

The American Gas Association's (AGA) laboratories on East 55th Street were approx. 600 feet (183 meters) south of No. 2 Works. There was a parking lot and open ground in-between, and Tank 4 was at a higher elevation; so they had a clear view of Tank 4. Witnesses to the beginning of Tank 4's failure reported that at 2:40 p.m., streams of liquid or vapor/fog issued from the side of the tank, and then almost immediately the tank opened up and dumped its entire contents—1.1 million gallons (4160 m^3)—of LNG over the plant area and the adjoining property [1,2,4].

The LNG spilled into an ambient atmosphere that was far above –250°F (–157°C), but with such a huge volume of LNG, evaporation did not happen immediately. The surface of the liquid could absorb heat from the ambient air and evaporate rapidly; but underneath this surface layer, the LNG remained a liquid. The LNG spread in its liquid form even as the surface layer was continually evaporating. The mass of LNG was large enough that it could spread quite far in its liquid form before evaporating. The LNG flowed out mainly in the east and southeast directions; it quickly engulfed the EOG laboratory, office, and meter shops [4].

Since the No. 2 Works were at a higher elevation, a large amount of the LNG flowed down 62nd Street and went down into the storm sewers.

The evaporating natural gas was highly combustible and there were ignition sources everywhere. Soon the No. 2 Works was in flames. The fire was so intense that the steel legs supporting Tank 3, the spherical tank nearest to Tank 4, collapsed from the heat [4] (see Figure 8.4). Another 2100 m^3 of LNG was released. The subsequent explosion shot flames half a mile into the air, and the temperature in some areas reached 3000°F (1649°C).

FIGURE 8.4
The collapsed Tank 3. Tanks 1 and 2 are visible behind it. In the foreground is the concrete foundation for Tank 4. (Courtesy of Cleveland Press Collection, Cleveland State University Library, Cleveland, OH.)

8.5.1 The Fires

The evaporating natural gas found ignition sources everywhere. Soon there were different types of fires going on:

- Pool fires from the flowing LNG.
- Explosions in sewers.
- Secondary fires ignited by the intense radiant heat, in areas not touched by the natural gas (either in liquid or vapor phase).
- Secondary fires ignited by rock wool, which had been soaked with LNG when Tank 4 collapsed and then flung into the air when Tank 3 exploded.
- The coal pile south of Tank 2 was burning.

8.5.1.1 Pool Fires

If a large mass of LNG spills, the liquid flows while its surface is evaporating. The evaporating gas is highly combustible; if there is any ignition source, then it starts burning above the LNG pool. The resulting "pool fire" spreads as the LNG pool expands and keeps evaporating. A pool fire generates intense heat, burning more hotly and rapidly than oil or gasoline fires [11], and may injure people and damage property a considerable distance from the fire itself.

Pool fires end only when all the natural gas is consumed.

8.5.1.2 Spreading through Sewer Explosions

The LNG entered storm sewers and main intercept sewers, where it evaporated. In the confined spaces of the sewer system, the result was explosions. A series of explosions occurred in the sewers and in the basements of nearby houses and commercial buildings. Streets were blown up, manhole covers were hurled into the air, and water lines were broken. One such explosion caused a crater 30 feet wide × 60 feet long × 25 feet deep (9 × 18 × 7½ meters), which swallowed a fire department pumper truck. Smaller blasts continued for hours.

As the fire swept through the sewers, it spread to houses connected to the sewer system. In Figure 8.5 one can almost visualize the progress of the conflagration through the sewer under this street. The rightmost house has just been engulfed and is burning from the inside out; fires are more advanced in houses to the left.

8.5.1.3 Intense Radiant Heat and Secondary Fires

The amount of energy stored in Tanks 3 and 4 was enormous (see Table 8.1). The ensuing fire was intensely hot, reaching up to 3000°F (1649°C) in spots.

FIGURE 8.5
Houses burning from inside out as fire spreads through sewers. (Courtesy of Cleveland Press Collection, Cleveland State University Library, Cleveland, OH.)

TABLE 8.1

Energy Released by Tanks 3 and 4

	Capacity, LNG Converted to Gas		Equivalent Energy		
Tank	Cubic Feet	Cubic Meters	Btu	Therm	Megajoule
3	50,000,000	1,415,850	51.4×10^9	51,400	54,229,870
4	100,000,000	2,831,700	102.8×10^9	102,800	108,459,741

Assumption: 1 ft³ = 1028 Btu (2014 average heat content of natural gas in the United States) [14].

The heat was sufficient to start secondary fires in areas not touched by the LNG or the vaporized natural gas. The buildings on Lake Court NE and the New York Central Railroad bridge were almost certainly ignited by radiant heat; the Bureau of Mines [4] found no evidence that any quantity of burning liquid came into this area.

The fire at Lake Court was extremely worrying. Adjacent to this residential area was the Ohio Chemical & Manufacturing Co. works. In order to stop the fire from overtaking this massive chemical plant, a U.S. Coast Guard fireboat on Lake Erie pumped water through 1400 feet (427 meters) of hoses to Lake Court on the northwest edge of the fire. There the Guardsmen managed to check the conflagration before it reached the adjacent chemical works (see Figure 8.6) [3].

FIGURE 8.6
U.S. Coast Guard fighting flames on Lake Court. (Courtesy of Cleveland Press Collection, Cleveland State University Library, Cleveland, OH.)

8.5.2 Map of the Area

The heavy dashed line in Figure 8.7 shows the extent of the main fire. (See Table 8.2 for explanation of numbered items in Figure 8.7.) There was widespread damage beyond the area enclosed by the dashed line (and beyond what is shown in the figure), but the heaviest damage occurred here.

TABLE 8.2

Key to Numbered Items in Figure 8.7

No.	Description
1	LNG Tank 1
2	LNG Tank 2
3	LNG Tank 3
4	LNG Tank 4
5	Former gas holder, used for LNG overflow
6	Waterseal gas holder
7	Houses and Ohio Chemical & Manufacturing Co. works saved by U.S. Coast Guard
8	N.Y. Central Railroad bridge, damaged by fire
9	American Gas Association Laboratory
10	Parking lot
11	Crater that swallowed a fire department pumper truck

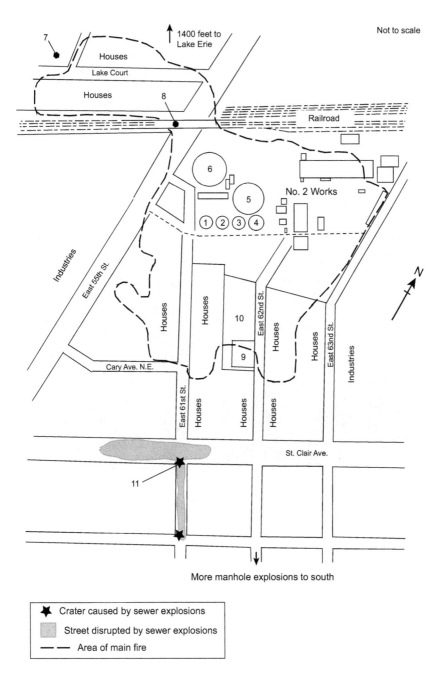

FIGURE 8.7
Extent of main fire and craters caused by sewer explosions.

8.6 The Days After

The affected area was about 0.5 sq. mile (1.3 km³). About 30 acres were completely devastated; everything combustible had burned [8] (see Figure 8.8).

The coroner's final tally was 130 people dead and 225 injured [12]. The greatest loss of life was within the No. 2 Plant, where 73 employees were killed.

The Cleveland fire chief reported to the Mayor's Board on the number of houses, factories, and vehicles destroyed in the fire (see Table 8.3).

FIGURE 8.8
Searching for victims. (Photo by Byron Filkins; Courtesy of Cleveland Press Collection, Cleveland State University Library, Cleveland, OH.)

TABLE 8.3

Houses, Factories, and Vehicles Destroyed in Fire

	Totally Destroyed	Partially Destroyed
Houses	79	35
Factories	2	13
Automobiles	217	
Trailers	7	
Tractor	1	

Source: Report of the Technical Consultants Board of Inquiry for the Mayor of Cleveland into the East Ohio Gas Company Fire, Cleveland, OH, July 1945.

On Saturday and Sunday, the main body of fire was under control but there were still fires burning:

- Vent gas from Tanks 1 and 2 was burning at a break in the vent gas line.
- A fire was burning in a line, probably a former vent gas line, near the collapsed Tank 3.
- The pile of coal south of Tank 2 was still burning.
- Smoke was coming out of the vent at the top of Tank 2 (separately from the break in the vent line).

The tank designer, J.O. Jackson, climbed to the top of Tank 2 on Sunday morning to investigate and found that the smoke was coming from burning cork insulation. Jackson arranged with the fire department to get as much liquid and solid carbon dioxide as possible to the scene, and the fire in the cork insulation was smothered.

8.6.1 Emptying Tanks 1 and 2

The rehabilitation work included emptying Tanks 1 and 2. Even though they had withstood the fire on Friday, and the cork insulation fire was extinguished, Tanks 1 and 2 presented a problem. They each contained 50 million cubic feet (1.4 million cubic meters) of natural gas condensed to liquid form. However, the regasification plant that could safely turn the LNG into gas had just been destroyed in the fire. How, then, to dispose of the LNG without starting another conflagration?

Gas company employees arranged piping from the vent lines of the two spheres to a point about 350 feet (107 meters) from Tank 2 (see Figure 8.9). Emergency steam lines were run into the plant from locomotives on the tracks of the New York Central Railroad (see Figure 8.10). The steam was used to carefully vaporize the LNG stored in Tanks 1 and 2. The process took several weeks. When the tanks were finally empty, they were purged with inert gas.

8.6.2 Rock Wool Everywhere

The rehabilitation work and the investigations were complicated by the rock wool that had been used to insulate Tank 4. The explosion/fire had sent one million pounds of rock wool airborne, and when it came down it was distributed over a large part of the No. 2 Plant area. Fragments of tank and plant equipment were covered by up to two feet of rock wool. The gas company finally had to arrange for the removal of the mineral wool to help find fragments of Tank 4 [4].

8.6.3 Investigations Launched

There were at least nine investigations launched after the explosion and fire, including a probe launched by the mayor of Cleveland (the "Mayor's Board"), an investigation by the Coroner for Cuyahoga County, and one by the National

FIGURE 8.9
East Ohio Gas Co. workmen Geo. Spicer, Pat Joyce, Mike McGinty, and Pat Malloy hook up steam pipe running to Tanks 1 and 2. (Photo by Tom Brunton; Courtesy of Cleveland Press Collection, Cleveland State University Library, Cleveland, OH.)

Board of Fire Underwriters [1]. By far the most important investigation, however, was that carried out by the federal government's Bureau of Mines.

8.7 Investigations by the Bureau of Mines and the Mayor's Board of Inquiry

The Bureau of Mines and the Mayor's Board cooperated closely on their investigations.

The Act of Congress that created the Bureau of Mines also charged that bureau with responsibilities for investigating the causes of accidents and making recommendations for preventing future occurrences. The bureau brought to their investigation a great deal of experience, expert knowledge, and resources for failure analysis.

FIGURE 8.10
Engine supplying steam to Tanks 1 and 2. The remaining LNG tanks can be seen in the background. (Photo by Walter Kneal; Courtesy of Cleveland Press Collection, Cleveland State University Library, Cleveland, OH.)

8.7.1 Observations

No direct evidence of the events immediately prior to the failure of Tank 4 was available, since almost everybody close to the LS&R plant was killed. The survivors could offer only piecemeal facts that, when put together, did not offer many clues about the cause.

Therefore, a large part of the failure analysis investigation was to gather and examine as many remnants as possible of Tank 4.

8.7.1.1 Did Tank 4 Explode or Disintegrate?

The indications are that the cylindrical tank probably disintegrated rather than exploded. The experts from the Bureau of Mines [4] came to this conclusion, based upon

- The type of failure
- The size of fragments
- The geographical distribution of fragments. The detailed fragment map created under the supervision of the Mayor's Board showed that no fragments were found at a great distance from the site.

Four large sections of the bottom ring of the tank, weighing between one and two tons (900–1800 kg) each, were found 200–300 feet (61–91 meters) from the tank. The force needed to move these weights over such distances is well within the range of force that could be supplied by the sudden release of hydrostatic pressure when the tank failed. It is not necessary to presume an explosion, in order to account for these pieces moving this distance.

8.7.1.2 Location on Tank 4 of Initial Rupture

The damaged outer shell (the insulation jacket) of Tank 4 was found as more or less one continuous sheet of metal after the explosion. Its position and condition fixed the location of the initial breakthrough of LNG to the east or southeast side. There were also indications that the major portion of the LNG was released in this direction [4].

8.7.1.3 Analysis of Tank 4 Inner Shell Steel

A sample of the inner shell of Tank 4 was analyzed by the Bureau of Mines metallurgy laboratories in Pittsburgh, to see if the steel provided conformed to the specifications of the tank designers. As Table 8.4 shows, the steel met the specification ranges for constituents, as set by the Pittsburgh-Des Moines Steel Co.

The Bureau of Mines did find, in examining fragments of Tank 4, typical embrittlement fractures [4].

8.7.2 Possible Failure Scenarios

Several possible scenarios were considered by the Bureau of Mines for the failure of Tank 4 [4]. These are briefly summarized in Table 8.5.

8.7.3 Recommendations and Conclusions from the Investigations

In their 1945 report, the Bureau of Mines made 17 recommendations. Today, 70 years later, several of them are so well-established industry practice that

TABLE 8.4

Tank 4 Inner Shell Steel Specification and Analysis of Steel Fragment

Constituent	Specification Range, Given by Pittsburgh-Des Moines Steel Co. (Percent by Weight)	Sample of Inner Shell, Tank 4 (Percent by Weight)
Carbon	0.08–0.12	0.12
Manganese	0.30–0.60	0.40
Sulfur	0.045 maximum	0.19
Phosphorus	0.04 maximum	0.025
Silicon	0.10–0.20	0.14
Nickel	3.25–3.75	3.45

Source: Elliott, M. A. et al., Report on the investigation of the fire at the liquefaction, storage, and regasification plant of the East Ohio Gas Co., Cleveland Ohio, October 20, 1944, Report of investigations, United States Department of the Interior, Bureau of Mines, Bulletin 3867, Washington, DC, 1946.

TABLE 8.5

Failure Scenarios Considered by the Bureau of Mines

Scenario	Observations
Event external to Tank 4 damages tank, causes gas leakage, which ignites.	No evidence to support this.
Explosive shock from burst ammonia cylinder.	Cylinder burst did occur, based on fragments found. Location of fragments indicates this was after Tank 4 failed.
Shock from sudden pressure release, due to broken vent line.	Witness accounts do not support this. If the broken vent line preceded the Tank 4 failure, then witnesses at AGA would have seen a cloud of vapor between Tanks 2 and 3.
Shock from failure in liquefaction plant.	Charts show rise in pressure; however, this could have been caused by the fire. Tank 4 was off-line; difficult to see how shock could be transmitted.
Seismic shocks.	Not proven nor disproven. • Known that minor shocks can cause failure of brittle materials when stressed. • Vibrations known to exist, e.g., from trains and drop-hammer at nearby factory. • Concrete foundation connected to inner shell through wood posts. Vibrations in concrete pad could be transmitted to inner shell.
Crack, strain, or flaw in metal.	Possible that a minute crack passed unnoticed and that the region near the crack gradually weakened and failed. • Temperature differences due to settling insulation were considered. However, it was believed that the temperature differences in the shell would not be great enough to cause the strain needed for failure.
Superheating of LNG in Tank 4.	The liquid level gages gave no indication of this.

they almost cause raised eyebrows: for example, fire and damage control drills, explosion proofing all electrical lines and equipment in such plants, and reacting swiftly if frost forms on LNG tanks. Another recommendation is to study the properties of metals at low temperatures; metallurgists have certainly made great advances in this since 1944.

Three recommendations are extremely pertinent today [4]:

1. Plants dealing with large quantities of liquefied flammable gases should be isolated at considerable distance from inhabited areas.

2. Containers for the liquefied gas should be isolated from other parts of the plant and provided with dikes large enough to contain the entire contents of the tank.

3. Extreme precaution should be taken to prevent spilled liquefied gas from entering storm sewers or other underground conduits.

The Mayor's Board made a number of recommendations, several of which were covered by the Bureau of Mines' list. Among their recommendations that were not duplicated in the other report are the following [13]:

- An application for construction should include a full statement of the energy content and maximum possible rate of energy dissipation.
- The owner should be required to prove that he offers no hazard to the surroundings and that firefighting and emergency provisions are adequate.

8.8 Summary

The disaster of October 20, 1944, killed 130 people, injured over 200 more, destroyed two factories, and burned out a neighborhood in the city of Cleveland. It also more or less killed interest in LNG for the next 15 years or so.

The staff and management of the EOG Co. who operated the facility, and the Pittsburgh-Des Moines Steel Co. who designed the tank, were trained, experienced, and reputable. They were perhaps unlucky in the timing of the world's first commercial LNG plant, before material science could provide them with adequate knowledge of all the potential problems.

Quite a lot has changed since 1944. Metallurgy has come a long way: the ductile-to-brittle transition curve and its implications, which eluded engineers in the 1930s, is common knowledge among material science students today. Fatigue stressing of metals and welds at extremely low temperatures is better understood. Steels used for these applications are now 9% nickel alloy, not 3½% [8]. Tanks today are insulated with perlite or polyurethane foam, not rock wool or cork [8]. And the formation of frost spots on the outside of LNG tanks today would cause alarm bells and swift remedial action.

The cylindrical shape of Tank 4 came in for a lot of criticism at the time; some reports written soon after the event demand that no more such tanks be built, until they could be proven safe. However, there is no reason to believe that the choice of a *cylindrical shape* for Tank 4 caused the accident. All large LNG tanks today are cylindrical [8]. As mentioned earlier, today's steel alloy is different.

Those who have investigated this disaster—the Bureau of Mines and the Mayor's Board of Inquiry at the time, or the Comptroller General, reporting to Congress in the 1970s [4,8,13]—came to conclusions that are still highly relevant:

1. Plants dealing with large quantities of LNG should not be located in residential areas.
2. Extreme caution should be taken to prevent spilled liquefied gas from entering storm sewers.

References

1. Bellamy, J. S. (2009). *Cleveland's Greatest Disasters*. Gray & Company Publishers: Cleveland, OH, pp. 21–40.
2. Lemoff, T. (February 15, 2010). Early LNG experience in the US. In: *SCCP Workshop on Safe Storage of Natural Gases*. U.S.-India Standards and Conformance Cooperation Programme (SCCP): New Delhi, India.
3. Sandy, E. (October 15, 2014). The Day Cleveland Exploded: 70 Years Later, the Unthinkable Disaster of the East Ohio Gas Co. Explosion. *Cleveland Scene*. Available at: http://www.clevescene.com. Accessed December 28, 2015.
4. Elliott, M. A., Seibel, C. W., Brown, F. W., Artz, R. T., and Berger, L. B. (1946). Report on the investigation of the fire at the liquefaction, storage, and regasification plant of the East Ohio Gas Co., Cleveland Ohio, October 20, 1944. Report of investigations: United States Department of the Interior, Bureau of Mines, Bulletin 3867: Washington, DC. UNT Digital Library. http://digital.library.unt.edu/ark:/67531/metadc38535/. Accessed January 4, 2016.
5. DOE. (August 2005). Liquefied natural gas: Understanding the basic facts. Publication No. DOE/FE-0489. U.S. Department of Energy, Office of Fossil Energy: Washington, DC.
6. Clark, J. A. and Miller, R. W. (October 17, 1940). Liquefaction, storage, and regasification of natural gas. *Oil and Gas Journal*, 50–52.
7. Miller, R. W. and Clark, J. A. (January 1941). Liquefying natural gas for peak-load supply. *Chemical and Metallurgical Engineering*, 48(1), 74–76.
8. Staats, E. B. (July 31, 1978). Report to the Congress by the Comptroller General: Liquefied energy gases safety, 3 vols. Publ. EMD 78-28. United States General Accounting Office: Washington, DC.
9. Foley v. the Pittsburgh-Des Moines Company. Supreme Court of Pennsylvania, 363 Pa. 1(1949). Foley, Executrix, Appellant, v. The Pittsburgh-Des Moines Company et al. Opinion by Mr. Justice Horace Stern, September 26, 1949.
10. NFPA. (1943). *National Fire Codes for Flammable Liquids, Gases, Chemicals and Explosives*. Boston, MA: National Fire Protection Association (NFPA).
11. SCCP. (February 15, 2010). Background paper: Interactive session on US-India cooperation in standards & conformance in LNG, LPG & CNG equipment. In: *SCCP Workshop on Safe Storage of Natural Gases*. U.S.-India Standards and Conformance Cooperation Programme (SCCP): New Delhi, India.
12. Coroner's Report on East Ohio Gas Company Disaster, Cuyahoga County, Cleveland, Ohio. (July 1945). Conclusion and recommendations from the Coroner's Report are included in the "Report of the Technical Consultants Board of Inquiry for the Mayor of Cleveland into the East Ohio Gas Company Fire."
13. Barnes, G. E., Braidech, M. M., and Donaldson, K. H. (1945). Report of the Technical Consultants Board of Inquiry for the Mayor of Cleveland into the East Ohio Gas Company Fire. Board of Inquiry on East Ohio Gas Company Fire: Cleveland, OH.
14. Energy Information Administration, eia.gov/tools/FAQS. Accessed January 24, 2016.

9

Other Vapors or Gases

9.1 Carbon Monoxide

Carbon monoxide (CO) is a highly flammable, highly toxic gas, created by incomplete combustion.* CO is one of the most common sources of poisoning deaths in the world [1,2].

CO is the most common cause of fatal gas inhalation exposures in the U.S. workplace, accounting for 36% of the deaths. Estimates of the annual number of unintentional deaths in the United States due to CO poisoning range from 300 to 900[†] [3–6]. In 1997, CO poisoning caused 108 unintentional deaths in England and Wales [7].

9.1.1 Properties

Table 9.1 shows some important physical and chemical properties of CO.

CO is a flammability hazard—the flammability range is quite extensive, 12.5%–74%. However, its toxicity is what kills.

The density of this gas is very close to that of air, so it can disperse fairly easily. Pockets of CO can form, however, in areas with little air movement.

9.1.1.1 Other Names

CO is known as white damp, coal gas, and flue gas.

9.1.2 How Carbon Monoxide Kills

CO is a chemical asphyxiant. It prevents oxygen from reaching the tissues of the body in sufficient quantities to sustain life. Normally, the protein hemoglobin in the red blood cells takes up oxygen (forming oxyhemoglobin) and distributes it throughout the body. CO interferes with this by competing with oxygen for the binding sites on the hemoglobin. Approximately 95% of

---

* Plants can also produce CO as a metabolic by-product [5] but not in the amounts needed to cause CO poisoning.
† The lower numbers reflect non-fire-related CO poisonings.

TABLE 9.1

Properties of Carbon Monoxide

CAS number	630-08-0
Density (air = 1)	0.9678
Color	Colorless
Odor	Odorless
Flammability	LFL = 12,500 ppm, or 12.5% v/v in air
	UFL = 74,000 ppm or 74%
Toxicity	Toxic; NIOSH Immediately Dangerous to Life and Health (IDLH) level = 1200 ppm

the CO binds to the hemoglobin to form carboxyhemoglobin (COHb), which is useless for distributing O_2 around the body. Tissue hypoxia results. The tissue hypoxia has the greatest effect on tissues that extract high percentages of the available oxygen, such as the brain and heart [7–9].

CO has a higher affinity than O_2 to the hemoglobin in the red blood corpuscles. It outcompetes O_2 for the binding sites on hemoglobin by at least 200:1, so even small amounts of CO create a hazard [7,9–11].

The COHb is formed quickly, but it takes a significant amount of time to leave the blood (see Figure 9.1). The half-life of COHb in the blood is reported to be four to five hours in air [7].

By binding to hemoglobin (Hb), the CO directly diminishes the oxygen-carrying capacity of the blood. There is a second effect also, due to COHb: what little oxygen there is in the blood tends to stay there rather than being made available to tissues. The presence of COHb alters the equilibrium of the remaining oxyhemoglobin (HbO_2), driving it away from O_2–Hb dissociation. The blood's ability to unload O_2 at the tissue is lowered, further reducing tissue partial pressure of oxygen (Pa_{O_2}) [5,12,13].

Recent research has indicated that CO may also have another toxic effect: inhibiting the activity of cytochrome a3 oxidase [5,13–15], which is needed for cellular energy production (see Section 4.2.3.2). As Dydek [5] notes, "these latter findings help to explain the clinical experience that carboxyhemoglobin (COHb) levels (an indicator of the risk of tissue hypoxia) are a very poor predictor of a patient's medical condition and his or her prognosis."

9.1.3 Exposure Effects

Exposure effects are shown in Table 9.2; they can range from subtle to severe and depend upon the concentration and exposure time. In general, at low levels, CO produces neurobehavioral, cardiovascular, and developmental effects; at high concentrations, CO leads to unconsciousness and death [9].

Overall, lethal poisonings tend to be linked to exposures of 1000 ppm or greater. COHb levels are more difficult to interpret: when COHb levels reach 50%, coma, convulsions, cardiopulmonary arrest, and death are the frequent result.

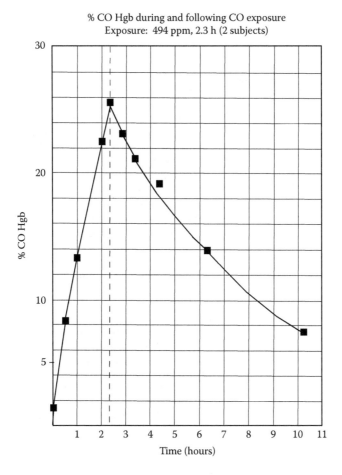

FIGURE 9.1
Absorption and excretion of carbon monoxide (494 ppm breathed for 2.3 hours). (Modified from Stewart, R.D. et al., *Arch. Environ. Health Int. J.*, 21(2), 154, 1970.)

However, COHb below 50% should not be interpreted as "safe" since many deaths from CO poisoning occur at much lower COHb levels [9,12].

In Table 9.2, the studies exposing volunteers to approximately 100 ppm measured very different levels of COHb in the blood [20,21,23,24]. This may be due to differing measuring techniques, especially given the broad time span. It may also reflect a difference in the volunteer groups. Schulte's study tested middle-aged firefighters, almost all of whom were smokers; Stewart et al. tested medical students and medical school faculty, aged 24–42, almost none of whom were smokers; and Allred et al. studied three large groups of nonsmokers who suffered from coronary artery disease.

Also, as Schulte [19] has pointed out, the way that the elevated COHb level is reached (i.e., slow ramping up of CO in the inhaled air, or a sudden massive

TABLE 9.2

Exposure Effects of Carbon Monoxide

CO Level			
PPM in Air	% COHb in Blood	Effects	Reference
1–3	0.8–0.7	Normal.	[16]
25	NR	Nausea, sleepiness, and headaches.	[17]
NR	5.1%	Decreased maximum exercise duration; decreased maximum effort. No effect on heart rate or rhythm, cardiac perfusion, or blood pressure.	[18]
30–60	5–10	Exercise tolerance is reduced.	[16]
NR	ca 10%	Impaired driving ability. Drivers could not maintain the distance to car in front.	[19]
100	Up to 20%	Adverse effect on cognitive abilities and choice discriminations.	[20]
100	11%–13%	Volunteers exposed to 100 ppm for 8 hours showed no impairment in coordination, reaction time, or manual dexterity tests.	[21,22]
117	2.0%	4.2% decrease in time to onset of angina during exercise (study of 63 adult nonsmokers with coronary artery disease).	[23,24]
NR	5%–20%	Visual and auditory sensory effects; changes in fine and sensorimotor performance; cognitive effects; brain electrical activity.	[25]
60–150	10%–20%	Frontal headache; shortness of breath on exertion.	[16]
NR	10%–20%	Headaches.	[27]
200	16%	Headache.	[21]
253	3.9%	7.1% decrease in time to onset of angina during exercise (study of 63 adult nonsmokers with coronary artery disease).	[23,24]
150–300	20–30	Throbbing headache, dizziness, nausea, manual dexterity impaired.	[16]
NR	20–30	Headache, throbbing in temples.	[26]
NR	35%	Impaired manual dexterity.	[28]
NR	30%–40%	Severe headache, weakness, dizziness, dimness of vision, nausea and vomiting, collapse.	[26]
NR	40%	Mental confusion, increased incoordination.	[28]
300–650	30%–50%	Severe headache, nausea and vomiting, confusion, and collapse.	[16]

(Continued)

TABLE 9.2 (*Continued*)

Exposure Effects of Carbon Monoxide

CO Level			
PPM in Air	**% COHb in Blood**	**Effects**	**Reference**
494	25.5%	Headache; nausea. Walking to blood sampling port caused 10% increase in heart rate.	[21]
NR	40%–60%	Acute and delayed onset neurological impairment (ranging from headache to coma) and pathology.	[29]
700–1000	50%–65%	Coma, convulsions.	[16]
1000	32%	Incapacitating severe headache, lasting through following day.	[21]
1000–2000	65%–70%	Heart and lung function impaired, fatal if not treated.	[16]
NR	60%–70%	Coma with intermittent convulsions, depressed heart action and respiration, possible death.	[26]
Over 2000	Over 70%	Unconsciousness and death.	[16]
NR	70%–80%	Weak pulse and slowed respiration; respiratory failure and death.	[26]

Abbreviation: NR, not reported.

"hit" of CO) might easily have an effect on test results. There is some indication of this in animal trials; canine studies have suggested that CO diffusing into tissues during long exposures have a cumulative effect [12].

The emphasis on smokers versus nonsmokers is simply because smoking generates CO and therefore affects the background COHb. Romieu [30] has reported that average COHb levels are 1.2%–1.5% in the general population and 3%–4% in cigarette smokers.

9.1.3.1 Neuropsychological Sequelae

Massive overexposure to CO can cause permanent damage, usually to the nervous system. The range of symptoms and sequelae is great, almost bewilderingly so. Possibly this is because the mechanisms of neurotoxicity are many, complex, and multifactorial [5,13,31,32].

CO toxicity can also lead to delayed sequelae: that is, after a period of apparent recovery following CO poisoning, a wide spectrum of neurological, behavioral, and psychiatric disturbances is manifested. Delayed neurological syndrome usually appears within four to six weeks of the CO exposure, but much longer delays have been reported in the literature. The mechanisms for delayed neurological syndrome are far from understood [5,7,14,32].

Permanent effects can include, among other things, the following [5,16,32–34]:

- Memory problems
- Mental deficits/deterioration/learning disabilities
- Increased irritability
- Impulsiveness/mood changes
- Personality changes
- Violent behavior/verbal aggression
- Instability when walking
- Decreased motor skills
- Sleep disturbances
- Vision problems
- Hearing loss

Chambers et al. [33] have found that CO-related cognitive sequelae, depression, and anxiety are common and seem to be independent of poisoning severity.

The number of people estimated to sustain long-term cognitive sequelae every year in the United States is staggering: an estimated 14% of CO-poisoning survivors or approximately 6600 cases [6,14].

9.1.3.2 Cardiovascular Sequelae

The heart is an organ with very high oxygen requirements, so it is not surprising that cardiovascular sequelae are often seen after CO poisoning. Sequelae include sinus tachycardia, atrial flutter and fibrillation, premature ventricular contractions, ventricular tachycardia and fibrillation, coronary hypoperfusion, cardiac dysrhythmia, ischemia, and myocardial infarction. CO can exacerbate underlying cardiovascular disease and cause cardiac injury even in persons with normal coronary arteries [14,15,34].

Satran et al. [35] studied 230 cases of moderate-to-severe CO poisoning. The patients were not chosen because of any cardiovascular symptoms; instead, they examined 230 consecutive cases treated in the hyperbaric oxygen chamber at a regional center for CO poisoning. They found that cardiovascular sequelae are frequent in such cases, with myocardial injury in 85 patients (37%). In a follow-up study, they found a substantially increased long-term mortality among the 85 patients who had acute myocardial injuries: 38% of the group had died at a median follow-up of 7.6 years. Of the 145 patients who had not sustained myocardial injury, the comparable mortality rate was 15% [36].

Workers in the field recommend that patients admitted to the hospital with CO poisoning should have a cardiovascular investigation, including baseline ECG [34–36].

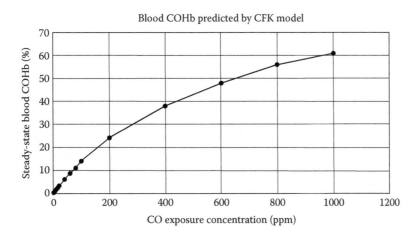

FIGURE 9.2
COHb levels in blood, predicted by Coburn–Forster–Kane model. (Data from Agency for Toxic Substances and Disease Registry (ATSDR), Toxicological profile for carbon monoxide, U.S. Department of Health and Human Services, Public Health Service, Atlanta, GA, June 2012 [online], available at http://www.atsdr.cdc.gov/toxprofiles/tp201.pdf, accessed December 20, 2015.)

9.1.3.3 Exposure Concentration versus COHb Levels

Figure 9.2 shows the steady-state blood COHb levels predicted by the Coburn–Forster–Kane model, for inhaled concentrations of CO between 0.1 and 1000 ppm [29].

It should be noted that this graph is based on a model, and a wide person-to-person variance should be expected.

9.1.3.4 Endogenous Carbon Monoxide

An interesting note is that not all the COHb is necessarily due to inhaled CO. The human body produces minuscule amounts of endogenous CO.

Molecules of CO, like hydrogen sulfide, have been identified as a gaseous signaling compound. This is endogenous CO, generated by the body in minuscule amounts to be used as a gasotransmitter (see Section 3.5.2).

Another endogenous source of CO is the body's normal breakdown and replacement of red blood cells; this produces low amounts of COHb [7,11,15,37].

Some solvents, such as methylene chloride (dichloromethane), can metabolically degrade in the blood to produce, among other chemicals, CO, thereby elevating COHb. Some drugs are also reported to have this effect [9,15].

We should repeat here that the levels of CO produced by endogenous sources are very low and not enough to cause symptoms. CO does not reach toxic concentrations unless it is inhaled from exogenous sources [7].

TABLE 9.3

Carbon Monoxide Exposure Limits. (Unless otherwise noted TWA is for eight hours and STEL is for 15 minutes.)

Jurisdiction	Organization	Notes	Limit
United States	OSHA		TWA: 50 ppm
United States	NIOSH	Nonregulatory; advisory body to OSHA	TWA: 35 ppm Ceiling: 200 ppm
Australia	Safe Work Australia		TWA: 30 ppm
Canada—Ontario	Ontario Ministry of Labour		TWA: 25 ppm STEL: 100 ppm
United Kingdom	HSE		TWA: 30 ppm STEL: 200 ppm
N/A	American Conference of Governmental Industrial Hygienists (ACGIH)	Nongovernmental organization	TWA: 25 ppm

9.1.4 Exposure Limits

Exposure limits for CO are shown in Table 9.3.

9.1.5 Sources of CO in Sewers and Sewer Projects

CO is produced by the incomplete combustion of carbon-containing solids, liquids, and gases. During complete combustion, water, carbon dioxide, and other oxides are formed; under incomplete combustion, CO is produced as well.

The most common sources of CO for our industry are

1. Diesel or gasoline engines
2. Exhaust gases from explosive blasting

Welding and cutting operations can also produce CO and should always be done with adequate exhaust venting.

9.1.5.1 Internal Combustion Engines

The most common source of CO is diesel or gasoline engines. The exhaust gases of internal combustion engines can contain large amounts of CO. If these engines are used in a confined space or inside a building, then deadly levels of CO can develop.

Even using these engines outside the confined space, but near the entrance, can be dangerous, since exhaust gases can find their way into the space. If stationary engines or fuel-powered vehicles must be used near the entry point of the confined space, then precautions must be taken to ensure that exhaust fumes are directed away from the opening.

9.1.5.2 Explosive Blasting

CO is one of the exhaust gases produced by explosive blasting. Because it can migrate through soil after surface blasting, it sometimes causes fatalities on sewer-building projects. Migration of CO through soil after surface blasting was first reported during World War I, where tunneling and blasting were done in an attempt to penetrate enemy lines. The phenomenon of CO migration through soil is an area that might benefit from more attention [37].

In a 1997 incident, CO migrating through soil after the nearby use of explosives caused three cases of CO poisoning, one of which was fatal [37]. On the day of the accident, a construction crew working on a municipal sewer project installed a 12-foot deep manhole and then left the area. The manhole was not yet connected to any sewer or water lines. After they left the area, 265 pounds of nitroglycerin-based explosive was detonated in a surface blast approximately 40–60 feet from the manhole. The explosive was in 22 boreholes, each 18 feet deep. The surface blast was done to prepare the ground for future pipe-laying operations.

Forty-five minutes after the explosion, a worker entered the manhole and collapsed within minutes. Two coworkers descended into the manhole to rescue him. One rescuer managed to bring the unconscious worker to the surface and then collapsed there. The other rescuer tried to climb out of the manhole but lost consciousness. He was eventually brought out by a fireman wearing self-contained breathing apparatus but was declared dead upon arrival at the hospital. All three workers had elevated levels of COHb (23%–33%), showing that they had inhaled high CO concentrations. Sampling of the manhole air approximately six feet down from the surface (the limit of the probe's length) showed 600 ppm of CO and 12 ppm of H_2S.

The subsequent investigation determined that CO from the explosion had migrated through the soil into the sewer manhole. Two days after the accident, the CO levels at the bottom of the manhole were 1905 ppm (compared to the IDLH level of 1200 ppm). The sewer manhole was ventilated, but high levels of CO reappeared because of continued diffusion from the surrounding soil. The CO diffusing in tapered off until only 6 ppm was measured eight days later.

There were multiple entries into the manhole that day; confined space entry precautions, including air monitoring, were not utilized before any of them [37,38].

Auger et al. [39] describe an accident involving CO generated by explosives. In this case, two adults living in a house without any obvious sources of combustion gases were found to suffer from CO poisoning. The severity was sufficient that they had to be treated in a hyperbaric chamber (HBO treatment).

CO, methane, and nitrogen oxide levels were monitored in the house air; CO readings were also taken in neighboring houses. Soil samples were collected around the house and tested for hydrocarbon residues.

A pocket of CO was found under the foundation of the house. CO levels in the basement were initially 500 ppm. The CO contamination then tapered off over a week. The investigation found that the probable source of the CO was the use of explosives at a nearby rain sewer construction site.

9.2 Carbon Dioxide

Carbon dioxide (CO_2) is not flammable or explosive. It is a colorless, odorless gas that is normally present in the atmosphere in small amounts, approx. 0.03–0.38 volume % [40–44].

CO_2 is produced by animal respiration, plant decay, and the burning of fossil fuels. It is also produced by bacterial breakdown of hydrocarbons under aerobic conditions. If there is a petroleum source in the soil, such as gasoline or diesel fuel from a spill, then the bacterial degradation can cause a buildup of CO_2 in the soil. CO_2 levels measured in soils in the vicinity of subsurface petroleum contaminations have been as high as 30% [45].

Environments with high levels of CO_2 are a health hazard in two ways:

1. Toxicity: At concentrations greater than 10% (100,000 ppm), CO_2 causes physiological effects that can lead to respiratory and cardiac arrest.
2. Oxygen deficiency: Large amounts of CO_2 can displace air and reduce the amount of available oxygen, causing asphyxia.

The two effects are quite separate. In air, if the CO_2 level rises to 50% v/v, then the O_2 concentration is reduced to a level that is immediately dangerous to human life. However, a CO_2 level of 15% poses an immediate threat to life due to CO_2 toxic effects, even if the O_2 concentration remains at 21% [41,46].

9.2.1 Physical and Chemical Properties

Table 9.4 shows some important physical and chemical properties of CO_2.

The density of CO_2—about 1½ times that of air—means that it can accumulate in low-lying areas and confined spaces, creating a health hazard.

9.2.1.1 Other Names

CO_2 is known as carbonic acid gas, blackdamp, stythe, and dry ice (in solid form) [43,44].

TABLE 9.4

Properties of Carbon Dioxide

CAS number	124-38-9
Density (g/L)	1.98
Color	Colorless
Odor	Odorless
Flammability	Noncombustible
Toxicity	Toxic at high concentrations
	NIOSH IDLH level = 40,000 ppm
Solubility	Very soluble in water

9.2.2 Exposure Effects

CO_2 is a physical asphyxiant; if the CO_2 concentration is high enough, then it displaces O_2 and causes anoxia [41,47]. It is also toxic at high concentrations. Target organs are the respiratory system and cardiovascular system.

CO_2 can induce the following [40,42,44]:

- Headache
- Dizziness
- Restlessness
- Sweating
- Increased heart rate, cardiac output, and blood pressure
- Visual disturbances
- Malaise (vague feeling of discomfort)
- Reduced ability to reason
- Dyspnea (breathing difficulty)
- Unconsciousness, coma
- Convulsions
- Asphyxia

At high concentrations, CO_2 is toxic; the clinical symptoms it induces when the concentration exceeds that found in the atmosphere are shown in Table 9.5. The seriousness of the symptoms depends on both the concentration of CO_2 and the length of exposure; response can also vary greatly in healthy individuals. The symptoms may be of short or long duration [43].

CO_2 is denser than air, so it collects in low points and local depressions. This can make it very dangerous when excavating for sewer lines.

At very high concentrations, for example, over 15% CO_2, several sources attribute death to hypoxia rather than direct toxic effects. As Harper et al.

TABLE 9.5

Exposure Effects of Carbon Dioxide

Concentration (Vol.%)	Symptoms	Reference
Low level	Tachycardia, hyperventilation, headache	[47]
Low level	Labored breathing, drowsiness, headache	[48]
2.5%	Increased ventilation	[42]
3%	Headaches	[46]
3%–6%	Headache, shortness of breath	[41]
5%	Increased ventilation, no activity restrictions within the exposure limit	[42]
	Exposure limit: 8 hours	
5%	Signs of intoxication after 30 minutes	[49]
5%–7%	Cerebral vasodilation, general vasoconstriction	
7%	Increased ventilation; severe limitations in activity	[42]
	Exposure limit: <30 minutes	
7%–10%	Unconsciousness in a few minutes	[49]
10%	Loss of consciousness	[51]
10%	Increased heart rate; collapse/unconsciousness	[42]
	Exposure limit: <2.0 minutes	
10%–11%	Visual distortion, headache, tremors, rapid loss of consciousness	[41]
10% or more	Cannot be endured for more than 10 minutes, since it acts on the nerves of the respiratory system	[52]
15%	Unconsciousness, coma, and death	[46]
22%	Death likely	[41]
Higher concentrations	Confusion, loss of consciousness, death due to hypoxia	[43,47]
Higher concentrations	Death by suffocation	[48]

[46] point out, the precise cause of death probably does not matter: "At CO_2 concentrations in excess of 50% in air whether a person dies due to the toxicological effect of CO_2 inhalation or due to oxygen depletion is not clear and arguably immaterial. In both cases death would be the outcome."

9.2.2.1 The Role CO_2 Plays in Breathing

CO_2 is a normal by-product of cellular respiration. It is transported from the body's cells to the alveoli in the lungs, where it is exhaled to the atmosphere.

In the human body, CO_2 plays an important role in the control of breathing; it is known to be one of the most powerful stimuli causing a speeding up or slowing down of breathing. This makes humans very sensitive to changes in CO_2 concentrations; the sensitivity is even more marked above levels of about 7% CO_2 in air [46].

9.2.2.2 *CO₂ and Circulation*

Kety and Schmidt [50] exposed volunteers to 5% or 7% CO_2 (while O_2 was maintained at 21%). Their experiments point to CO_2 causing cerebral vasodilation and vasoconstriction in the rest of the body.

They found that the cerebral blood flow increased 75%. This was accompanied by a reduction in the mean cerebrovascular resistance (from 1.6 to 1.1 resistance units). Cerebral oxygen consumption was not significantly changed.

For the general circulatory effects of CO_2, they report a significant rise in arterial blood pressure, though they note that cardiac output was not significantly changed. They interpret this as indicating a net peripheral vasoconstriction.

9.2.3 Exposure Limits

Exposure limits for CO_2 are shown in Table 9.6.

9.2.3.1 *Concentration versus Time*

The UK Health and Safety Executive has estimated CO_2 exposure concentrations and durations that have a significant likelihood of death (SLOD). The SLOD is defined as causing 50% lethality from a single exposure, given a certain exposure time. The calculations were developed from routine toxicity testing on animals. The resulting SLODs are shown in Table 9.7.

TABLE 9.6

Carbon Dioxide Exposure Limits. (Unless otherwise noted, TWA is for 8 hours and STEL is for 15 minutes.)

Jurisdiction	Organization	Notes	Limit
United States	OSHA		TWA: 5,000 ppm
	NIOSH	Nonregulatory; advisory body to OSHA	TWA: 5,000 ppm; STEL: 30,000 ppm
Australia	Safe Work Australia	The Work Health and Safety (WHS) Act permits a higher TWA value in coal mines	TWA: 5,000 ppm; STEL: 30,000 ppm
Canada	CCOHS		Uses the ACGIH TWA and STEL-C
United Kingdom	HSE		TWA: 5,000 ppm; STEL: 15,000 ppm
N/A	ACGIH	Nongovernmental organization	TWA: 5,000 ppm; STEL-C: 30,000 ppm

TABLE 9.7

Significant Likelihood of Death Consequences of Concentration of Carbon Dioxide in Air versus Time

Inhalation Exposure Time (Minutes)	Inhaled % CO_2 Leading to SLOD: 50% Fatalities (%)	Inhaled ppm CO_2 Leading to SLOD: 50% Fatalities (ppm)
60	8.4	84,000
30	9.2	92,000
20	9.6	96,000
10	10.5	105,000
5	11.5	115,000
1	14	140,000

Source: Harper, P., et al., Assessment of the major hazard potential of carbon dioxide (CO_2), Health and Safety Executive, Liverpool, UK, June 1–28, 2011.

Note: Contains public sector information published by the Health and Safety Executive and licensed under the Open Government Licence.

9.2.4 Sources of Carbon Dioxide

Levels of CO_2 can exceed the amounts found in the normal atmosphere, due to various natural and anthropogenic causes such as

- Combustion of organic material
- Decomposition of organic material
- Acid waters, such as acidic rainwaters, reacting on carbonate rocks, such as chalk or limestone [41,43]
- Respiration

9.2.4.1 Carbon Dioxide's Association with Methane

CO_2 is often produced alongside methane in the decomposition of organic material. The processes that produce coal, or landfill gas, also produce CO_2. In naturally occurring sources of methane, therefore, CO_2 is often present.

9.3 Other Chemicals

The number of organic and inorganic chemicals used in a modern society is almost unfathomable; and almost any of them can appear in wastewater. As infrastructure ages, leaks from other utilities can cause contamination in sewer lines. Unauthorized discharge by industries into sewer lines is a constant hazard. Private households sometimes dispose of oil, paint, and solvents by flushing them into the sewer, whether or not this is legal.

Even small amounts of such chemicals can present fire and explosion dangers, because they volatilize in the sewer system. In the confined space of a sewer, a seemingly insignificant amount is enough to form a potentially explosive atmosphere.

Chemicals can legitimately find their way into the system, too: for example, traces of oil can easily be washed from roads into the storm water sewers.

It is not possible to discuss the potential effects of all the chemicals that might turn up in wastewater or even a significant fraction of the chemicals. The presence in significant concentration of any particular potentially toxic compound may require a site-specific evaluation of the potential health effects.

In this section, we will have to be content with briefly discussing gasoline and ammonia and then giving an illustrative example of the havoc an unauthorized discharge caused in the midwestern United States in 1977.

9.3.1 Gasoline (Petrol)

Gasoline (petrol) is a complex chemical mixture often including 250 separate hydrocarbons; some of them, for example, benzene, toluene, xylenes, and n-hexane, have well-established toxicity.

Gasoline can be found in both sanitary and storm sewer systems, as a result of illegal disposal, spills, or leakage from underground storage tanks. The April 22, 1992, explosions in Guadalajara, Mexico, which killed over 200 people, were due to gasoline leaking into the sewer system.

Gasoline is highly flammable; the vapors present a fire and explosion hazard. Here are some of the characteristics that make it such a fire hazard:

- It can be ignited under almost all ambient temperature conditions.
- It can float on water and spread fire.
- It can accumulate static charge.
- Vapors may travel to a distant source of ignition and then flash back to the source.

Gasoline vapor can cause headache, nausea, drowsiness, confusion, and dizziness. Severe exposure can cause unconsciousness. It is also a respiratory irritant, affecting the nose and throat. Gasoline liquid defats the skin; chronic skin contact can cause dry, red cracked skin (dermatitis) [17,53,54].

9.3.1.1 Physical and Chemical Properties

Table 9.8 shows some important physical and chemical properties of gasoline.

Because gasoline vapors are heavier than air, it can accumulate in low-lying areas, especially low-lying confined spaces.

TABLE 9.8

Properties of Gasoline

CAS numbers	8006-61-9
	86290-81-5
Density (g/L)	Liquid: 0.7–0.8 (water = 1.0).
	Vapors are 3–4 times heavier than air.
Color	Colorless to amber colored.
Odor	Petroleum-like.
Flammability	Very flammable liquid and vapor. Distant ignition and flashback are possible.
Explosive limits	1.3%–7.1% v/v [54].
Toxicity	Contains hydrocarbons of established toxicity.
Solubility	Insoluble in water.

9.3.2 Ammonia

Ammonia (NH_3) is a strongly alkaline gas created by the natural decay of organic matter. Exposure to elevated levels can severely irritate the eyes, nose, throat, and lungs. Repeated exposure to ammonia vapor may cause chronic irritation of the eyes and upper respiratory tract. Exposure to high concentrations can be fatal [48,55–57]. NH_3 has a good warning characteristic: a sharp, penetrating, pungent odor that is detected well below toxic levels.

The deposit that accumulates in sewers can be a source of NH_3 [58]. A study performed in Dundee, Scotland, indicates that old sediments in particular can act as a reservoir of ammoniacal nitrogen, which can be released to the water during peak flows [59–61]. The ammonia dissolved in the water presents environmental challenges; but whether it will become airborne, in sufficient quantities to create a health hazard, is an area that could profit from more research.

In the farming industry there has been concern about potential health effects of ammonia, because it exists in significant quantities, for example, in livestock houses, waste storage/treatment systems, and manure application sites.

In our industry, it is not likely that acutely toxic levels of NH_3 would be present in common municipal sewer systems. NH_3 is associated more with wastewater treatment plants (WWTP), or drinking water treatment plants, than sewer systems. This is dependent upon the sewer system, of course, and factors such as dilution with storm water/industrial water, or climate. Abeliovich and Azov [62], for instance, reported high levels of ammonia in the domestic raw sewage of Jerusalem: 4–8 mM NH_3 (0.068–0.136 g/L NH_3); Hodgson [63] reported a much lower level for Accra, Ghana: 0.018 g/L NH_3 on average.

Table 9.9 shows some important physical and chemical properties of ammonia, and Table 9.10 gives exposure limits.

TABLE 9.9

Properties of Ammonia

CAS number	7664-41-7
Specific gravity	0.77 at 32°F (0°C)
Vapor density	0.6 (air = 1 at boiling point of ammonia)
Boiling point (at 760 mm Hg)	−28°F (−33.4°C)
Color	Colorless
Odor	Characteristic, penetrating and pungent odor
Flammability	NFPA flammability rating: 1 (slight fire hazard); LFL = 16% v/v; UFL = 25% v/v
Toxicity	Very toxic; fatal if inhaled
Corrosive	Yes—strongly alkaline
Solubility	Very soluble in water, alcohol, ether, and chloroform

TABLE 9.10

NH_3 Exposure Limits. (Unless otherwise noted, TWA is for 8 hours and STEL is for 15 minutes.)

Jurisdiction	Organization	Notes	Limit
United States	OSHA		TWA: 50 ppm
United States	NIOSH	Nonregulatory; advisory body to OSHA	TWA: 25 ppm for a 10-hour period STEL: 35 ppm

9.3.3 Example of an Unauthorized Discharge

It has been estimated that there are 30,000 chemicals in commercial use in the United States. Morse et al. [64] and Kominsky [65] have described the havoc wrecked when one of them is discharged into the sewers in significant amounts.

In the middle of March 1977, workers at the Morris Forman Wastewater Treatment Plant in Louisville, Kentucky, began reporting unusual odors. On March 26, four employees used steam in an attempt to remove an odoriferous, highly viscous, and sticky substance from the bar screens and grit collection systems. This produced a blue haze that spread through the primary treatment area. The blue haze was anything but benign: it caused 20 workers to seek medical treatment for tracheobronchial irritation.

The next day, following a heavy rain, employees noticed a blue haze hovering over the grit collection channels and a disagreeable odor throughout the primary treatment area. On March 29, the WWTP had to be closed, after tests showed that the wastewater was contaminated with hexachlorocyclopentadiene (HCCPD) and octachlorocyclopentene. HCCPD is an intermediate compound in the manufacture of pesticides; it is also extremely toxic by dermal, oral, and inhalation routes of exposure.

The airborne HCCPD concentrations of the blue haze generated by the cleanup attempt were estimated to be over 19,000 ppb. (For comparison, the ACGIH's TLV (eight-hour TWA) for HCCPD in 1977 was 10 ppb.)

This unauthorized discharge of HCCPD into a municipal sewer line had far-reaching effects: workers at the Morris Forman WWTP were endangered; the treatment plant and contributory sewer lines had to be decontaminated; and cities as far as 200 miles away were affected. The Morris Forman WWTP discharges to the Ohio River. The EPA advised several cities downstream— Evansville, Indiana; Henderson, Kentucky; Mt. Vernon, Indiana; and Golconda, Illinois—drawing water from the Ohio River, to add activated carbon to their maximum capacity, until the potential danger period passed.

Workers at the WWTP suffered symptoms consistent with the known toxic properties of HCCPD: headache and irritation of skin, respiratory tract, and mucous membranes.

Two months afterward, at the end of May 1977, air measurements in the main interceptor sewer still showed high levels of HCCPD: 19 air samples had a range of 21–3833 ppb, with the mean 1528 ppb. Workers entering the sewer had to wear respirators for months after the illegal chemical dumping.

References

1. Braubach, M., Alqoet, A., Beaton, M., Laurios, S., Héroux, M. E., and Krzyzanowski, M. (2013). Mortality associated with exposure to carbon monoxide in WHO European member states. *Indoor Air*, 23, 115–125.
2. Oh, S. and Choi, S. C. (2015). Acute carbon monoxide poisoning and delayed neurological sequelae: A potential neuroprotection bundle therapy. *Neural Regeneration Research*, 10(1), 36.
3. Mott J. A., Woolfe, M. I., Alverson C. J., Macdonald, S. C., Bailey, C. R., Ball, L. B., Moorman, J. E., Somers, J. H., Mannino, D. M., and Redd, S. C. (2002). National vehicle emissions policies and declining US carbon monoxide-related mortality. *Journal of the American Medical Association*, 288(8), 988–995.
4. CDC. (2007). Carbon monoxide-related deaths: United States 1999–2004. *Morbidity and Mortality Weekly Report*, 56(50), 1309–1312. Centers for Disease Control and Prevention, Department of Health and Human Services: Washington, DC.
5. Dydek, T. M. (2008). Investigating carbon monoxide poisonings, in *Carbon Monoxide Poisoning*. CRC Press/Taylor & Francis Group: Boca Raton, FL, pp. 287–304.
6. Hampson, N. B. (2016). Cost of accidental carbon monoxide poisoning: A preventable expense. *Preventive Medicine Reports*, 3, 21–24.
7. McParland, M. (1999). Carbon monoxide poisoning. *Emergency Nurse*, 7(6), 18–22.

8. Adams, J. D., Erickson, H. H., and Stone, H. L. (September 9–11, 1970). The effects of carbon monoxide on coronary hemodynamics and left ventricular function in the conscious dog. In: *Proceedings of the First Annual Conference on Environmental Toxicology.* AMRL-TR-70-102. Wright-Patterson Air Force Base: Dayton, OH, pp. 107–110.

9. Flachsbart, P. G. (2008). Exposure to ambient and microenvironmental concentrations of carbon monoxide, in *Carbon Monoxide Poisoning*, ed. D. G. Penney. CRC Press LLC/Taylor & Francis Group: Boca Raton, FL, pp. 5–42.

10. Ilano, A. L. and Raffin, T. A. (1990). Management of carbon monoxide poisoning. *CHEST Journal*, 97(1), 165–169.

11. Prockop, L. D. and Chichkova, R. I. (2007). Carbon monoxide intoxication: An updated review. *Journal of the Neurological Sciences*, 262(1), 122–130.

12. Goldbaum, L. R., Orellano, T., and Dergal, E. (1976). Mechanism of the toxic action of carbon monoxide. *Annals of Clinical & Laboratory Science*, 6(4), 372–376.

13. Hampson, N. B., Piantadosi, C. A., Thom, S. R., and Weaver, L. K. (2012). Practice recommendations in the diagnosis, management, and prevention of carbon monoxide poisoning. *American Journal of Respiratory and Critical Care Medicine*, 186(11), 1095–1101.

14. Raub, J. A., Mathieu-Nolf, M., Hampson, N. B., and Thom, S. R. (2000). Carbon monoxide poisoning—A public health perspective. *Toxicology*, 145(1), 1–14.

15. Kao, L. W. and Nanagas, K. A. (2004). Carbon monoxide poisoning. *Emergency Medicine Clinics of North America*, 22(4), 985–1018.

16. IAPA. (2008). Carbon monoxide in the workplace. Industrial Accident Prevention Association (IAPA): Toronto, Ontario, Canada.

17. *Sewer Entry Guidelines.* (2010). Workplace Health and Safety Bulletin CH037. Queen's Printer, Government of Alberta, Department of Employment and Immigration: Edmonton, Alberta, Canada.

18. Adir, Y., Merdler, A., Ben Haim, S., Front, A., Harduf, R., and Bitterman, H. (1999). Effects of exposure to low concentrations of carbon monoxide on exercise performance and myocardial perfusion in young healthy men. *Occupational and Environmental Medicine*, 56(8), 535–538.

19. Schulte, J. (September 9–11, 1970). The effect of exposure to low concentrations of carbon monoxide. In: *Proceedings of the First Annual Conference on Environmental Toxicology.* AMRL-TR-70-102. Wright-Patterson Air Force Base: Dayton, OH, pp. 87–91.

20. Schulte, J. H. (1963). Effect of mild carbon monoxide intoxication. *Archives of Environment Health*, 7, 523–530.

21. Stewart, R. D., Peterson, J. E., Hosko, M. J., Baretta, E. D., Dodd, H. C., Newton, P. E., Fisher, T. N., and Herrmann, A. A. (September 9–11, 1970). Experimental human exposure to carbon monoxide <1 to 1000 PPM. In: *Proceedings of the First Annual Conference on Environmental Toxicology.* AMRL-TR-70102. Wright-Patterson Air Force Base: Dayton, OH, pp. 49–76.

22. Stewart, R. D., Peterson, J. E., Baretta, E. D., Bachand, R. T., Hosko, M. J., and Herrmann, A. A. (1970). Experimental human exposure to carbon monoxide. *Archives of Environmental Health: An International Journal*, 21(2), 154–164.

23. Allred, E. N., Bleecker, E. R., Chaitman, B. R., Dahms, T. E., Gottlieb, S. O., Hackney, J. D., Pagano, M., Selvester, R. H., Walden, S. M., and Warren, J. (1989). Short-term effects of carbon monoxide exposure on the exercise performance of subjects with coronary artery disease. *New England Journal of Medicine*, 321(21), 1426–1432.

24. Allred, E. N., Bleecker, E. R., Chaitman, B. R., Dahms, T. E., Gottlieb, S. O., Hackney, J. D., Pagano, M., Selvester, R. H., Walden, S. M., and Warren, J. (1991). Effects of carbon monoxide on myocardial ischemia. *Environmental Health Perspectives*, 91, 89.

25. Benignus, V. A. (1994). Behavioral effects of carbon monoxide: Meta analyses and extrapolations. *Journal of Applied Physiology*, 76(3), 1310–1316.

26. Sayers, R. R. and Davenport, S. J. (1936). Review of carbon monoxide poisoning. Public Health Bulletin No. 195. United States Government Printing Office: Washington, DC.

27. National Institute for Occupational Safety and Health (NIOSH). (1972). NIOSH criteria for a recommended standard: Occupation exposure to carbon monoxide. Publication No. HSM 73-11000. U.S. Department of Health, Education and Welfare: Cincinnati, OH.

28. Stewart, R. D. (1975). The effect of carbon monoxide on humans. *Annual Review of Pharmacology*, 15(1), 409–423.

29. Agency for Toxic Substances and Disease Registry (ATSDR). (June 2012). Toxicological profile for carbon monoxide. U.S. Department of Health and Human Services, Public Health Service: Atlanta, GA. Available: http://www. atsdr.cdc.gov/toxprofiles/tp201.pdf. Accessed December 20, 2015.

30. Romieu, I. (1999). Epidemiological studies of health effects arising from motor vehicle air pollution, in *Urban Traffic Pollution*, eds. D. Schwela and O. Zali. E&FN Spon: London, UK, Chapter 2.

31. Victor, M. and Adams, R. D. (1980). Metabolic diseases of the nervous system, in *Harrison's Principles of Internal Medicine*, 9th ed., eds. K. J. Isselbacher, et al. McGraw-Hill Book Company: New York, pp. 1978–1979.

32. Hopkins, R. O. (2008). Neurocognitive and affective sequelae of carbon monoxide poisoning, in *Carbon Monoxide Poisoning.*, ed. D. G. Penney. CRC Press LLC/ Taylor & Francis Group: Boca Raton, FL, pp. 477–495.

33. Chambers, C. A., Hopkins, R. O., Weaver, L. K., and Key, C. (2008). Cognitive and affective outcomes of more severe compared to less severe carbon monoxide poisoning. *Brain Injury*, 22(5), 387–395.

34. Weaver, L. K. (2009). Carbon monoxide poisoning. *New England Journal of Medicine*, 360(12), 1217–1225.

35. Satran, D., Henry, C. R., Adkinson, C., Nicholson, C. I., Bracha, Y., and Henry, T. D. (2005). Cardiovascular manifestations of moderate to severe carbon monoxide poisoning. *Journal of the American College of Cardiology*, 45(9), 1513–1516.

36. Henry, C. R., Satran, D., Lindgren, B., Adkinson, C., Nicholson, C. I., and Henry, T. D. (2006). Myocardial injury and long-term mortality following moderate to severe carbon monoxide poisoning. *Journal of the American Medical Association*, 295(4), 398–402.

37. Decker, J. A. et al. (1998). Department of Health and Human Services, Centers for Disease Control and Prevention, National Institute for Occupational Safety and Health. DHHS (NIOSH) Publication No. 98-122. U.S. Government Printing Office: Pittsburgh, PA.

38. Decker, J. A., Deitchman, S., and Santis, L. (1997). Carbon monoxide intoxication and death in a newly constructed Sewer Manhole. Report HETA 98-0020. Centers for Disease Control and Prevention, National Institute for Occupational Safety and Health: Atlanta, GA.

39. Auger, P. L, Lévesque, B., Martel, R., Prud'homme, H., Bellemare, D., Barbeau, C., Lachance, P., and Rhainds, M. (1999). An unusual case of carbon monoxide poisoning. *Environment Health Perspective*, 107(7), 603–605.

40. Farrar, C. D., Neil, J. M., and Howle, J. F. (1999). Magmatic carbon dioxide emissions at Mammoth Mountain, California. Water Resources Investigations report 98-4217. U.S. Geological Survey: Sacramento, CA.

41. NHBC. (March 2007). Guidance on evaluation of development proposals on sites where methane and carbon dioxide are present. Report No. 04. The National House-Building Council (NHBC): Amersham, UK.

42. Williams, J. W. (September 17, 2009). Physiological responses to oxygen and carbon dioxide in the breathing environment. In: Presentation from *NIOSH Public Meeting*, Pittsburgh, PA. National Institute for Occupational Safety and Health (NIOSH), Technology Research Branch: Pittsburgh, PA.

43. Appleton, J. D. (2011). User guide for the BGS methane and carbon dioxide from natural sources and coal mining dataset for Great Britain. BGS Open Report, OR/11/054. British Geological Survey: Keyworth, UK.

44. National Institute for Occupational Safety and Health (NIOSH). (2015). NIOSH pocket guide database: Carbon dioxide. Department of Health and Human Services, Centers for Disease Control and Prevention: Pittsburgh, PA. Available: http://www.cdc.gov/niosh/npg/npgd0103.html. Accessed December 28, 2015.

45. Jones, III, V. T. and Agostino, P. N. (November, 1998). Case studies of anaerobic methane generation at a variety of hydrocarbon fuel contaminated sites. In: *Chemicals in Ground Water: Prevention, Detection, and Remediation; Conference and Exposition*, pp. 11–13.

46. Harper, P., Wilday, J., and Bilio, M. (June 2011). Assessment of the major hazard potential of carbon dioxide (CO_2). Health and Safety Executive: Liverpool, UK, pp. 1–28.

47. Knight, L. D. and Presnell, S. E. (2005). Death by sewer gas: Case report of a double fatality and review of the literature. *The American Journal of Forensic Medicine and Pathology*, 26(2), 181–185.

48. National Institute for Occupational Safety and Health (NIOSH). (2007). Preventing deaths of farm workers in manure pits. NIOSH Publication No. 90-103. U.S. Department of Health and Human Services, Centers for Disease Control, National Institute for Occupational Safety and Health: Morgantown, WV.

49. National Institute for Occupational Safety and Health (NIOSH). (May 1994). NIOSH immediately dangerous to life or health concentration database: Carbon dioxide. Department of Health and Human Services, Centers for Disease Control and Prevention, National Institute for Occupational Safety and Health (NIOSH): Pittsburgh, PA. Available: http://www.cdc.gov/niosh/idlh/124389.html. Accessed December 28, 2015.

50. Kety, S. S. and Schmidt, C. F. (1948). The effects of altered arterial tensions of carbon dioxide and oxygen on cerebral blood flow and cerebral oxygen consumption of normal young men. *Journal of Clinical Investigation*, 27(4), 484.

51. Budavari, S., Dale, M., O'Neil, J., and Smith, A., eds. (1989). *Merck Index—Encyclopedia of Chemicals, Drugs, and Biologicals*, 11th ed. Rahway, NJ: Merck and Company.

52. Gray, C., Vaughn, J. L. and Sanger, K. (2011). *Distribution/Collection Certification Study Guide*. Oklahoma State Department of Environmental Quality: Oklahoma City, OK.

53. Canadian Centre for Occupational Health and Safety (CCOHS). OSH Fact Sheet, Gasoline. CCOHS: Hamilton, Ontario, Canada. Available: http://www.ccohs.ca/products/databases. Accessed December 28, 2015.

54. National Institute for Occupational Safety and Health (NIOSH). (2016). International chemical safety cards: Gasoline. ICSC 1400. Department of Health and Human Services, Centers for Disease Control and Prevention, National Institute for Occupational Safety and Health (NIOSH): Pittsburgh, PA. Available: http://www.cdc.gov/niosh/ipcsneng/neng1400.html. Accessed January 14, 2016.

55. Michaels, R. A. (1999). Emergency planning and the acute toxic potency of inhaled ammonia. *Environmental Health Perspectives*, 107(8), 617.

56. Agency for Toxic Substances and Disease Registry (ATSDR). (June 2004). Toxicological profile for ammonia. U.S. Department of Health and Human Services, Public Health Service, Agency for Toxic Substances and Disease Registry (ATSDR): Atlanta, GA. Available: http://www.atsdr.cdc.gov/toxprofiles/tp126.pdf. Accessed June 12, 2011.

57. Burton, N. C. and Dowell, C. (2011). Health hazard evaluation report: Evaluation of exposures associated with cleaning and maintaining composting toilets—Arizona. Publication NIOSH HETA No. 2009-0100-3135. U.S. Department of Health and Human Services, Centers for Disease Control and Prevention, National Institute for Occupational Safety and Health (NIOSH): Cincinnati, OH.

58. Saul, A. J. and Thornton, R. C. (1989). Hydraulic performance and control of pollutants discharged from a combined sewer storage overflow. In: *Urban Discharges and Receiving Water Quality Impacts, International Association on Water Pollution Research and Control 14th Biennial Conference*, Brighton, UK, pp. 113–122. Pergamon Press: Oxford, UK.

59. Ashley, R. M., Wotherspoon, D. J. J., Coghlan, B. P., and McGregor, I. (1992). The erosion and movement of sediments and associated pollutants in combined sewers. *Water Science Technology*, 25, 101–114.

60. McGregor, I. and Ashley, R. M. (1992). Specification of laboratory analysis and sampling of sewage and sewer sediments. Report for the Water Research Centre, U.K. Institute of Technology: Dundee, Scotland.

61. Dabrowski, W. (2000). Storage of ammonia in old sewer sediments. *Lakes & Reservoirs: Research & Management*, 5(2), 89–92.

62. Abeliovich, A. and Azov, Y. (1976). Toxicity of ammonia to algae in sewage oxidation ponds. *Applied and Environmental Microbiology*, 31(6), 801–806.

63. Hodgson, I. O. (2000). Treatment of domestic sewage at Akuse (Ghana). *Water S. A.*, 26(3), 413–415.

64. Morse, D. L., Kominsky, J. R., Wisserman, C. L., and Landrigan, P. J. (1979). Occupational exposure to hexachlorocyclopentadiene: How safe is sewage? *Journal of the American Medical Association*, 241, 217–279.

65. Kominsky, J. R. (1980). Nonviable contaminants from wastewater: Hexachlorocyclopentadiene contamination of a municipal wastewater treatment plant, in *Wastewater Aerosols and Disease: Proceeding of a Symposium*, September, 19–21, 1979, eds. H. Pahren and W. Jakubowski. EPA-600/9-80-028. Environmental Protection Agency, Health Effects Research Laboratory: Cincinnati, OH.

10

Biological Hazards

10.1 Introduction to Biological Hazards

Wastewater contains microbes because the human feces it carries is full of microbes, dead and alive. Estimates of the fraction of feces that is microbial flora are between 30% and 95% w/w [1–3]. Wastewater is likely to contain bacteria, viruses, fungi, and parasites (protozoan and worms). The exact flora and concentration will depend on several factors: size of the population served, season, types of infections, and number of infected people in the community served [4–6].

Most of the sewage microbes will not cause disease. The majority of illnesses seem to be relatively mild cases of gastroenteritis [7], unless there is an extraordinary contagion in the community, such as an outbreak of winter vomiting disease (norovirus). However, pathogens can exist in the sewage that cause potentially fatal diseases, such as leptospirosis (Weil's disease) and hepatitis. And for all illnesses there is a possibility of significant under-reporting of cases because there is often failure to recognize the link between illness and work [7].

10.1.1 Pathogen Presence Does Not Equal Contracting Disease

The list of pathogens that can exist in wastewater is long and intimidating—until one remembers that the presence of a pathogen does not necessarily lead to contracting the disease.

Resistance to infection is determined by multiple factors [5,8,9]:

- The quantity of a particular pathogen required to initiate an infection.
- The state of physical barriers (skin, mucous membranes, respiratory epithelium, and gastric acid). Breaks in the barriers due to surgery, wounds, disease, or environmental stresses increase the vulnerability to infection.

- The unique pathogen–host interaction, including
  - Pathogen virulence
  - Amount of inoculum available
  - Genetically determined immune response
  - Immunization history

Sorber and Sagik [9] note that the quantity of a pathogen needed to cause illness is an extremely difficult factor to assess. Cliver points out that the dilution factors in wastewater are tremendously high, in the range of 1,000–10,000. Even during an outbreak of a disease, the amount of dilution is quite high: for example, if up to 10% of a community are infected and shed the pathogen at a level of 10^8 organisms/g feces, then in raw sewage this translates into levels of 10^5–10^7/L wastewater [5].

10.1.2 How Microorganisms Enter the Body

Pathogens in wastewater enter the body through

1. Being breathed in, as aerosol or dust
2. Ingestion—hand-to-mouth contact during eating, drinking, and smoking or swallowing mucous containing trapped pathogens [4,7]
3. Skin contact—for example, splashing of wastewater onto cuts or open wounds; wiping the face with contaminated gloves; penetrating wounds incurred while working in the sewer or with contaminated equipment

10.1.3 Protecting the Worker

The hazard cannot be eliminated, since microorganisms are an inherent part of sewage. It may be possible at times to minimize exposure to sewage, for example, by using remote-controlled robotic cameras for sewer inspection [7]. But the lion's share of the maintenance work usually requires entering the sewer system.

10.1.3.1 What the Employer Can Do

Safe work practices, personal protective equipment (PPE), and good hygiene are the best ways to protect workers from exposures to disease. In particular, employers can give the worker [7,10]

- Training and education about the hazards of wastewater and sewage
- The right PPE (depending on the job, this can include waterproof/abrasion-resistant gloves, footwear, eye and respiratory protection, and face visors to protect against splashes)

- A place onsite with clean water or alternative (e.g., premoistened towelettes) for washing hands
- A place to wash and clean up after work
- Clean areas set aside for eating
- Cleaning facilities or services for clothing and equipment (segregating the clean and contaminated clothing, PPE, and equipment is vital)
- First-aid equipment for cleansing wounds and sterile, waterproof, adhesive dressings
- Vaccinations and booster shots as appropriate and an effective system for monitoring the general health of the staff

10.1.3.2 What the Worker Can Do

Actions the worker can take to protect themselves include the following [7,10]:

- *Good hygiene*: Wash the hands well with clean water and soap before eating or smoking. Always wash the hands and exposed skin areas after work.
- Do not touch any part of the face or the ears with gloves on. Do not touch with bare hands, unless you have just washed.
- Wear waterproof gloves when in the sewers, when handling wastewater or sludge, or when handling wastewater equipment such as pumps or screens.
- Always wear waterproof gloves if the hands are chapped, cut, burned or have a rash.
- Shower and change out of work clothes before leaving work.
- Do not keep soiled work clothes with other clothes.
- Report any injury or illness that may be work related right away.
- When seeking medical attention, be sure to tell the doctor you work in a sewage or wastewater treatment plant. That information will help the doctor know what to look for.

10.2 Wastewater Bioaerosols

Aerosols are very small droplets, ca. 0.01–50 µm, of liquid that are suspended in air [9]. When wastewater is flung into the air in tiny droplets, bacteria and viruses in the wastewater are also ejected into the air in the bioaerosol.

Most studies of bioaerosols in our industry are centered on the wastewater treatment plant (WWTP), since the aerating, trickling, and spraying activities there are natural sources of aerosols. Aerosols form in sewer systems due to

- Cleaning equipment, especially hydrojet cleaning
- Collapsing bubbles
- Splashing

People inhaling aerosols can become infected because of the following [5]:

1. Inhalation leads to a respiratory infection.
2. The pathogens become entrapped in mucus, which is then swallowed. Infection takes place in the digestive tract.

Cliver notes that with the exception of *Mycobacterium tuberculosis*, few of the pathogens commonly found in sewage cause infection in the lungs. He suggests instead that sewage aerosols should be viewed principally as a route for fecal–oral transmission of enteric pathogens [5].

10.2.1 Bioaerosols from Jet Cleaning

Pressurized water (hydrojet/waterjet) is often used for cleaning or unblocking sewers. Although clean water is used in the jetting equipment, the bioaerosol generated can pick up sewage pathogens via the surfaces cleaned. De Serres et al. [11] have provided a vivid description of working with the mist caused by the powerful streams of water:

> Sewer cleaning operations are mechanized in many countries and use a machine that propels water in pipes under high pressure to clear the silt. This operation produces large amounts of aerosols that come out through manholes. Generally, workers do not wear full face protection and they must look in the manhole to drive the machine. During this operation, their face is soaked by aerosols, and they experience cutaneous, respiratory, and mucosal, and oral contact with sewage fluids.

10.2.2 Droplets from Splashing and Collapsing Bubbles

Splashing—from maintenance activities, cleaning with hoses, or turbulent flow of water—is a source of aerosol droplets. (It is also, of course, a source of direct skin contact.)

Aerosol droplets are also formed by the collapse of bubbles. Bubbles can be generated in sewers by biological decay of the sewage. When the bubble breaks on the water surface, it ejects a droplet that can contain the wastewater viruses and bacteria into the air [4]. The process is shown in Figure 10.1.

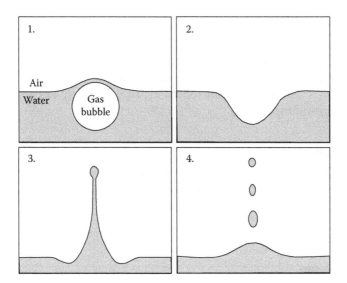

FIGURE 10.1
Bubble breaking and ejecting droplet.

10.2.2.1 Pathogen Concentration in Droplets

Interestingly, the density of a specific pathogen in the aerosol is not necessarily in proportion to the density of the organism in the wastewater. The pathogens are not necessarily evenly distributed throughout a liquid; instead there can be a bacteria-rich microlayer at the surface of the water. When air bubbles rise up to the air–water interface and break there, they may create droplets that are much richer in bacteria than the bulk of the wastewater liquid. Bacterial concentration in the newly formed droplets may be up to 1000 times that of the bulk wastewater, depending on drop size [4].

10.2.3 Survival of Microbes in Bioaerosols

Whether or not microbes in the bioaerosol will be able to cause disease in humans depends on several factors [4,9]:

- The fate of the pathogens in the sewage. Some pathogens, for example, the virus causing AIDS, simply do not survive in sewage.
- The amount of wastewater aerosolized and concentration of pathogens in it.
- The immediate effect of aerosolization on the organisms (aerosol shock).
- The biological decay of the organisms during transport as aerosols.

10.2.3.1 Aerosol Shock

Immediately after aerosolization, there is an initial die-off in many of the aerosolized organisms. The organisms' viability is impacted by environmental factors such as relative humidity, temperature, and sunlight. The factors leading to this initial die-off phenomenon are collectively known as aerosol shock or aerosol impact [4,9].

10.2.3.2 Biological Decay

After the aerosol shock, several factors contribute to the continued die-off or biological decay of the aerosolized organisms, for example, ultraviolet radiation, temperature, and relative humidity.

10.3 Bacteria

Pathogenic bacteria are very common in wastewater. Generally speaking, bacteria can reproduce outside the human body; some of them can be present in relatively large quantities in the sewer system. Fortunately, others, like *Shigella*, don't survive long in wastewater. Some of the bacteria of health concern that may be found in sewage are listed in Table 10.1.

10.3.1 Shigellosis

Shigellosis is an intestinal disease whose main sign is diarrhea (often bloody); it is the primary cause of infectious diarrhea in the United States [17]. Other common symptoms are abdominal pain, fever, nausea, and vomiting. Shigellosis can be transmitted through direct contact with the bacteria. A mild case usually clears up on its own in a week or so. Antibiotics are generally prescribed when treatment is needed.

Bacillary dysentery is a severe form of shigellosis, caused by one of several types of *Shigella* bacteria [18].

Shigella survives for only a short time in the sewer system [13].

10.3.2 Salmonellosis

Salmonellae live in the intestinal tracts of warm-blooded and cold-blooded animals. They cause acute gastroenteritis (infections of the stomach and intestinal tract), typhoid fever, and paratyphoid fever. Large numbers of the organisms are required to cause illness [13]. *Salmonella typhi*, which causes typhoid fever, has a relatively short survival time in wastewater [9].

TABLE 10.1

Selected Bacteria of Health Concern That May Be Present in Sewage

Organism	Disease	Transmission	Notes	Reference
Salmonella	Salmonellosis (gastroenteritis)	Ingestion	Approx. 1700 types. Large numbers of bacilli are needed to cause illness [13].	[4,12,13]
Salmonella typhi	Typhoid fever	Ingestion	Relatively short survival time in wastewater [9].	[9,13]
Shigella spp.	Shigellosis	Ingestion; inhalation	A very small dose (10–100 organisms) often suffices to cause illness. Survives for only a short time in the sewer system.	[12,13]
Enteric pathogens (*Yersinia enterocolitica; Escherichia coli*)	Gastroenteritis	Ingestion		[4,12,13]
Campylobacter fetus; Campylobacter jejuni	Bacterial enteritis	Ingestion	Most outbreaks are associated with surface waters and drinking water supplies.	[13]
Mycobacterium tuberculosis	Tuberculosis	Inhalation; ingestion		[9,13]
Vibrio cholerae	Asiatic cholera	Ingestion		[4,9,13]
Leptospira	Leptospirosis	Inhalation, wound contact, ingestion	See Chapter 12.	[13,14]
Clostridium tetani	Tetanus	Wound contact		[13,15]
Helicobacter pylori	Peptic ulcer disease			[16]

In the wastewater industry, salmonella infection is unlikely unless there has been direct ingestion [17].

10.3.3 Asiatic Cholera

The bacterium *Vibrio cholerae* produces an endotoxin that causes vomiting, diarrhea, and loss of body fluids [13]. It is transmitted by ingesting contaminated water.

10.3.4 Tetanus

When a deep wound or puncture wound is infected by *Clostridium tetani*, the disease tetanus can occur. The bacteria are commonly found in sewer systems and can therefore cause an infection whenever a deep or puncture wound is contaminated by wastewater.

Symptoms include

- Contraction of the muscles controlling the jaw
- Body muscle spasms
- Paralysis of throat muscle (causing death by respiratory failure)

The chances of encountering *C. tetani* are uncomfortably high in our industry; therefore, a vigilant tetanus vaccine program is highly recommended.

10.3.5 *Helicobacter pylori*

Helicobacter pylori is sometimes cited as a potential occupational hazard for wastewater workers. It causes peptic ulcer disease and has been identified as a risk factor for stomach cancer. The International Agency for Research on Cancer has classified this bacterium as a Class I human carcinogen.

Compared to other bacteria, not very much is known about *H. pylori*. Even the route of transmission is unclear [19], although fecal–oral, if not proven, seems reasonable.

Studies have not found an increased prevalence of peptic ulcer among sewage workers [16,19,20].

10.3.6 Endotoxins, Exotoxins, and Enterotoxins

Endotoxins are lipopolysaccharides; they are key structural components of the outer membrane of Gram-negative bacteria* (GNB). The lipopolysaccharides are often associated with the pathogenicity of the bacteria. Endotoxins are released by killed GNB as they decay. They can also be released by the activity of certain antibiotics or during bacterial cell growth. Endotoxins can cause fever, shock, blood pressure changes, and changes in other circulatory functions. All endotoxins produce a similar range of effects, regardless of the GNB source.

Exotoxins are protein toxins secreted by living bacteria, as part of their colonization of the host. In some cases, they can also be released by cell death. These protein toxins are specific to the bacteria type: for example, only *C. tetani*

---

* Most bacteria are classified into the groups Gram positive or Gram negative. The Gram stain test, developed by Hans Christian Gram in the 1800s, uses the difference in chemical structure of the bacterial cell wall to characterize the bacterium. A chemical stain is applied to color the bacteria's cell wall and then it is examined under a microscope. Gram-negative bacteria show a pink or red stain, and Gram-positive bacteria appear dark blue or purple.

produces the tetanus toxin. They can be very powerful poisons to humans: for example, the botulism, tetanus, and diphtheria toxins have toxicities that are orders of magnitude higher than that of strychnine. Both Gram-positive and Gram-negative bacteria can produce exotoxins. Generally, virulent strains of the bacterium produce the toxin.

Enterotoxins are a type of exotoxin, acting on the gastrointestinal tract. They produce typical food-poisoning symptoms. Lundholm and Rylander have suggested that enterotoxins from GNB may be responsible for the increased prevalence of gastrointestinal symptoms among sewage workers in Sweden [21].

10.3.6.1 Symptoms Caused by Endotoxins

Endotoxins are not infectious agents; but they are toxins and can cause ill-health. Endotoxin released from the cell walls of GNB after their death can produce fever and chest-tightness reactions in some exposed individuals [4,14].

Mattsby and Rylander [22] have reported that workers exposed to dried wastewater dusts had higher incidences of work-related fevers, diarrhea, and other symptoms consistent with endotoxin exposure.

More research is needed on airborne endotoxin exposure situations in sewers and WWTPs, to obtain relevant exposure determinations and establish preventive measures from a health risk perspective [23].

10.4 Viruses

Viruses, unlike bacteria, require a living host to reproduce. They cannot reproduce outside of a host cell and thus will not grow in wastewater. Human viruses sometimes found in wastewater are shown in Table 10.2.

There have been a number of studies of viral infections in sewage workers. The overall conclusion seems to be that *if workers are properly trained and properly protected*, then the actual risk of developing viral infections via occupational exposure is very low [14,15].

The category of virus that has been studied most in wastewater is the enteric (intestinal) virus [13]. Enteric viruses include the agents for infectious hepatitis, meningitis, and poliomyelitis.

10.4.1 HIV

The reader has no doubt already observed that HIV, the virus causing AIDS, is not listed in Table 10.2. The Centers for Disease Control and Prevention has stated that there is no scientific evidence that HIV is spread in wastewater or its aerosols.

TABLE 10.2

Selected Viruses of Health Concern That May Be Present in Sewage

Organism	Disease	Transmission	Notes	Reference
Adenoviruses	Diarrheal disease, gastroenteritis, respiratory disease, conjunctivitis, pharyngoconjunctival fever	Inhalation [13]	31 types.	[4,12,13]
Astrovirus	Gastroenteritis			[4]
Calicivirus	Gastroenteritis			[4]
Coxsackievirus A	Upper respiratory tract infection; hand, foot, and mouth disease; aseptic meningitis	Ingestion or contact; inhalation		[4,9,13]
Coxsackievirus B	Aseptic meningitis, upper respiratory tract infection, myocarditis	Ingestion or contact; inhalation		[4,9,13]
Echovirus	Aseptic meningitis, common cold, conjunctivitis, gastroenteritis	Inhalation [13]	The common cold is usually associated with rhinovirus types, coronaviruses, and some unknown viruses [13].	[4,9,13]
Enteroviruses	Gastroenteritis, heart anomalies, meningitis, others		67 types.	[4,12]
Hepatitis A virus	Hepatitis A	Ingestion [13]	See Chapter 11.	[4,12,13]
Hepatitis B virus	Hepatitis B		See Chapter 11.	[4]
Norwalk agent/ noroviruses	Gastroenteritis	Ingestion [13]; contact; possibly via airborne route	Only a few organisms are needed to cause illness.	[4,13]
Parvoviruses	Gastroenteritis			[4]
Parvovirus-like agents	Gastroenteritis		At least 2 types.	[12]
Polio 1, 2, 3	Poliomyelitis	Ingestion [13]		[4,9,13]
Reoviruses	Gastroenteritis, respiratory infection			[4,13]
Rotaviruses	Gastroenteritis	Ingestion [13]		[4,12,13]

HIV is a blood-borne virus: it cannot reproduce outside the human body. Merely coming into contact with wastewater does not imply exposure to AIDS. To be transmitted, HIV must enter the bloodstream directly. This would require both

1. A living virus in the wastewater
2. An open wound through which it can enter the bloodstream

10.4.1.1 Nonsurvival of HIV in Wastewater

HIV has never been recovered from wastewater [13]; it is believed that it has only limited survival in wastewater systems.

In sewer systems, HIV is subjected to environmental factors to which it is simply not suited: wrong pH, too low temperature, and the presence of surfactants and other chemicals [24]. Its die-off rate* in sewage is 10 times that of poliovirus [25].

The outer membrane of the virus seems to be fragile and does not hold up under conditions of unfavorable osmotic pressure. Such pressures can be caused by fluids that are different from those in the human body—such as wastewater. Without an intact outer envelope, HIV cannot infect [4].

10.4.1.2 Low Concentrations of HIV

The virus exists in the blood of infected persons in relatively low concentrations. By the time virus-contaminated fluids reach the sewer system, they are so enormously diluted that they do not present a significant risk to wastewater workers [13,15].

10.4.1.3 Needles

A more feasible risk scenario for becoming infected with HIV is following a puncture wound from a needle used by an HIV carrier; still, the risk for this is extremely low [15]. The virus seems to die off at temperatures cooler than the human body; and it appears to be vulnerable outside the body. Even in blood specimens kept at temperatures close to body temperature, the AIDS virus does not flourish [4].

The existing hygienic practices for dealing with sewage or removing debris from sewage equipment, which were designed to protect workers from more commonplace illnesses (e.g., hepatitis), seem sufficient to protect against AIDS as well [15].

---

* Measured as the time required for a 10-fold decrease in concentration.

10.4.1.4 No Reported Cases

To date, there are no reported cases of plumbers or wastewater workers who have contracted HIV/AIDS through occupational exposure [10,13].

10.4.2 Norwalk Agent (Norovirus)

Norwalk agent (norovirus) causes viral gastroenteritis characterized by vomiting, diarrhea, low-grade fever, and body aches. The noroviruses were first identified in an outbreak of vomiting and diarrheas in a primary school in Norwalk, Ohio (United States), in the 1970s. It is sometimes referred to as "winter vomiting disease."

Symptoms usually last for 24–48 hours; during this time the virus can be transmitted through the stool. Outbreaks of Norwalk agent have been associated with disposal of septic tank contents, municipal water supplies, and water-related recreational activities [13].

Occupational exposure to aerosolized noroviruses has been demonstrated by Uhrbrand et al. [26]; in a study of bioaerosols in a Danish WWTP, noroviruses were found on the dust filter worn by a worker.

10.4.3 Coxsackieviruses

Coxsackieviruses belong to the enterovirus group (which also includes polioviruses, echoviruses, and enteroviruses). There are two groups of coxsackieviruses, A and B; both can cause upper respiratory disease and can lead to aseptic meningitis.

Coxsackie group A viruses tend to infect the mucous membranes and the skin. They can cause the common cold, conjunctivitis, and the common viral illness hand, foot, and mouth disease [27].

Coxsackie group B viruses infect several organs (heart, pancreas, liver, pleura) and cause several types of diseases, for example, hepatitis, pleurodynia, and myocarditis [13].

Divizia et al. [28] have investigated seropositivity for Coxsackievirus B3 and Coxsackievirus B2 among sewer workers and a control group with no sewage exposure. They found a significant association between Coxsackievirus exposure and sewage work.

Coxsackieviruses can be checkmated by good hygiene.

10.4.4 Poliovirus

The poliovirus causes poliomyelitis, a disease affecting the central nervous system. This virus is more stable than most and can remain infectious for relatively long periods in contaminated water [13]. Polio vaccines have drastically reduced the incidence of polio in the general population and therefore the amount of this virus in the sewers.

10.5 Fungi

Disease-causing fungi that are often found in sewage are shown in Table 10.3.

10.5.1 Aspergillosis

Aspergillosis is a fungus-related illness caused by *Aspergillus* spp. Individuals whose immune systems are compromised, or who have asthma, are believed to be the most at risk for this infection. Aspergillosis can be fatal. People with a history of asthma can develop allergic bronchopulmonary aspergillosis [10,14].

10.5.2 Histoplasmosis

Histoplasmosis is an infection caused by the fungus *Histoplasma*. This is an opportunistic infection, most commonly found among people who are immunocompromised or are exposed to a large quantity of inoculum [29,30].

The fungus is associated with bird droppings or bat droppings, so sewer systems with bat infestations may have *Histoplasma*. Santos et al. [29] have reported on just such a case: three sewer workers were admitted to hospital for histoplasmosis, one of whom died. The authors make the point that the presence of bats in the sewer is often unknown, and therefore workers may not be using National Institute for Occupational Safety and Health–approved respirators when they encounter *Histoplasma*.

The *Histoplasma* spores are breathed in; most people do not get sick, but those who do may have a fever, cough, and fatigue [30]. One study has estimated that in the United States, 60%–90% of the people living in the Ohio River or Mississippi River valleys are exposed to the fungus at some point in their lifetime; however, most people exposed to the fungus do not contract the disease [31]. It is also endemic in Panama [32] and has been reported from San Luis Potosi state in Mexico [29].

TABLE 10.3

Selected Fungi of Health Concern That May Be Present in Sewage

Organism	Disease	Reference
Aspergillus spp.	Aspergillosis	[10,14]
Histoplasma	Histoplasmosis	[13,29]

10.6 Parasites

We frankly admit that parasites are included here under false pretenses. The scope of this book is airborne occupational hazards in sewer systems; bacteria, viruses, and fungi spores qualify because they can be found in bioaerosols. The case may be slightly different for, say, protozoa. But since it's difficult to talk about bacteria, viruses, and fungi in sewers without at least mentioning the protozoa and worms, we'll briefly touch on them.

A parasite lives on (or in) another organism, from a differing species. The parasite derives its nourishment from the host organism. Parasites typically do not kill their hosts, but they can cause illness and weaken them.

Hand-to-mouth contact is the principal cause of infection; good hygiene and frequent washing of hands is key to preventing parasitic infestation [13].

The parasites found in wastewater consist of protozoa and worms. The cysts and eggs of the protozoa and worms are often very hardy and still viable in the sewer system—indeed, many are viable even after passing through the WWTP [13].

10.6.1 Protozoa

The most commonly studied protozoa are *Entamoeba histolytica, Giardia lamblia,* and *Cryptosporidium* [10,13]. All three protozoa are contracted via ingestion. Some illness-causing protozoa that may be found in sewage are shown in Table 10.4.

E. histolytica causes amoebic dysentery, an acute enteritis ranging from mild diarrhea to fever and bloody and mucoidal diarrhea; in severe cases it can be fatal [14].

G. lamblia is a hardy protozoa that causes a variety of intestinal problems: giardiasis, chronic diarrhea, weight loss, and fatigue [10,14].

Cryptosporidium parasites are found throughout the world; in the United States alone, they cause an estimated 748,000 cases of the diarrheal disease cryptosporidiosis each year [33].

TABLE 10.4

Selected Protozoa of Health Concern That May Be Present in Sewage

Organism	Disease	Notes	Reference
Balantidium coli	Balantidiasis		[12]
Cryptosporidium	Cryptosporidiosis		[10]
Entamoeba histolytica, a/k/a *E. histolytica*	Amebiasis (a/k/a amoebic dysentery)		[4,12–14]
Giardia lamblia	Giardiasis	(Formerly called *Giardia intestinalis*)	[4,10,12–14]
Naegleria fowleri	Meningoencephalitis		[4]

TABLE 10.5

Worms Found in Wastewater

Organism	Disease
Roundworms (nematodes), including Ascaris	Abdominal pain, weight loss, ascariasis [12,13]
Hookworms	Anemia, hookworm disease [13]
Tapeworms (cestodes)	Abdominal pain, weight loss, taeniasis [12,13]

10.6.2 Helminths (Worms)

Table 10.5 shows the most common parasitic helminths (worms) found in wastewater.

Ascaris eggs are extremely hardy and have a relatively long survival time in wastewater [9,34]. They are sometimes used as indicator organisms of parasite contamination and to measure the survival of parasites in sludge [4].

Clark et al. have compared the prevalence of parasitic infections in 56 sewer workers to that of a control group of 69 highway maintenance workers in the midwestern United States. The sewer workers showed 5.4% infection, compared to 14.5% among the highway workers (engaged in cleaning). They conclude that parasitic infections do not occur more often among sewer workers, but that there may be a risk of parasitic infections among street/highway cleaners [35].

10.7 Sewage Worker's Syndrome

The title of this section is taken from a 1977 paper by Clark et al. [36]. A number of studies have investigated the health of sewage treatment workers and whether this is related to workplace exposures. Table 10.6 presents some such studies; this is only a fraction of the published literature, since a complete literature review is outside the scope of this chapter.

In surveys of the health of sewage workers, the impression is that gastrointestinal symptoms seem to dominate. This is perhaps not surprising since, as Brown points out, the disease-causing organisms found in sewage are mostly of enteric form, that is, organisms that infect the human intestinal tract. A minority of pathogens in sewage are known to cause respiratory infections [4].

The overall impression is that sewage workers tend to suffer more gastrointestinal and respiratory symptoms than their colleagues in other fields of municipal labor, such as drinking water or road maintenance workers.

TABLE 10.6

Selected Studies Investigating the Health of Sewage Treatment Workers

Location (Publication Date)	Comparison	Results	Notes	Reference
Cincinnati, United States (1977)	Three groups: Sewer workers, WWTP workers, and control group of road workers	Sewer workers had a significant increase in antibodies against enterovirus.	Age, gender, smoking habits not controlled for.	[36]
Cincinnati, Chicago, and Memphis, United States (1980)	Three groups: 78 sewer workers, 261 WWTP workers, and 185 controls	Found no consistent evidence of increased parasitic, bacterial, or viral infections in wastewater workers.	Newly employed sewage workers were more likely to have minor gastrointestinal illness than experienced workers.	[37]
Sweden (1983)	Workers at 6 sewage plants vs. workers at 3 drinking water plants	Statistically significant more skin disorders, diarrhea, and other gastrointestinal symptoms among the sewage plant workers.	Dominant flora in the bioaerosols was Gram-negative bacteria (GNB). Suggested toxins from GNB as cause of health effects.	[21]
Cincinnati, Chicago, and Memphis, United States (1985)	48 WWTP workers vs. control group of other municipal workers	No evidence found linking rotavirus or *Prototheca* to illness in WWTP workers. Inexperienced WWTP workers had higher levels of Norwalk antibody than did experienced workers or control group.	Study investigated Norwalk virus, rotavirus, and *Prototheca wickerhamii* prevalence in wastewater workers.	[38]
New York State, United States (1987)	189 sewage workers vs. control group of 82 drinking water workers	Sewage workers had significantly higher frequency of headache, dizziness, sore throat, skin irritation, and diarrhea.		[39]
Toronto, Canada (1988)	50 WWTP workers	Workers tended to have somewhat reduced lung function. Workers in sludge-drying area frequently reported an intermittent acute illness with cough, fever, and sore throat.	Basis for the possible health effects was not established.	[40]

(Continued)

TABLE 10.6 (Continued)

Selected Studies Investigating the Health of Sewage Treatment Workers

Location (Publication Date)	Comparison	Results	Notes	Reference
Zagreb, Croatia (1993)	Five groups: 26 closed-sewer workers, 31 open-drain sewage workers, 17 WWTP mechanics and vehicle drivers; 2 control groups, 35 bottling plant workers and 15 fruit juice workers	Both closed-sewer and open-drain sewage workers experienced significantly more chronic respiratory symptoms (cough, phlegm, bronchitis, chest tightness). Closed-sewer workers had highest prevalences of eye irritation, headache, dizziness, dyspnea, and burning or dryness of throat.		[41]
Mid-Sweden (1998)	142 sewage workers vs. control group of 137 other workers	No statistical difference between 2 groups in dyspepsia, diarrhea, irritable bowel syndrome. Numerically elevated risk for peptic ulcers among sewage workers.	Study examined abdominal symptoms; not general health of WWTP workers. Authors suggest peptic ulcer results may indicate that *Helicobacter pylori* is a risk factor for sewage workers and that more study is needed.	[20]
Netherlands (2001)	147 sewage workers at 51 WWTPs; chosen by management; no control group	Found significant relation between flu-like symptoms and working with sewage. Measured endotoxin exposure. Mean level was 9.5 EU/m^3 (85.6 EU/m^3 in sludge dewatering area).[a]	Concluded that endotoxins are not the cause of the health effects.	[42]
Norway (2010)	Three groups: 19 dry-sludge handlers; 25 other sewage workers; 36 office workers as control group	Dry-sludge handlers were exposed to higher levels of endotoxins and organic dust. Dry-sludge handlers had more airway and systemic symptoms. Correlation between symptoms and dust exposure was higher than correlation between symptoms and endotoxin exposure.	Study examined relation between sewage dust, endotoxins, and health; not general health of WWTP workers. Suggested organic dust, not endotoxins, as the cause of the health effects.	[43]

[a] In 1998, the Health Council of the Netherlands recommended 50 EU/m^3 as an occupation exposure limit [45].

But no clear picture emerges of one particular disease or group of diseases, or of one or two prevailing disease-causing agents. There seems to be something in "sewage worker's syndrome" but it is hard to say what.

We saw in Chapter 4 that hydrogen sulfide negatively affects the ability of pulmonary macrophages to mobilize and fight infection. It may be rewarding to get the toxicologists studying chronic low-level H_2S exposure in the same room as the epidemiologists studying the health of WWTP workers. Some, though far from all, of the health effects in Table 10.6 might be accounted for by low-level H_2S exposure.

For an excellent literature survey of studies examining the health of sewage treatment plant workers, we recommend the 2001 survey by Thorn and Kerekes [44].

References

1. Van Houte, J. and Gibbons, R. J. (1966). Studies of the cultivable flora of normal human feces. *Antonie van Leeuwenhoek*, 32(1), 212–222.
2. Moore, W. E. C. and Holdeman, L. V. (1975). Discussion of current bacteriological investigations of the relationships between intestinal flora, diet, and colon cancer. *Cancer Research*, 35(11 Part 2), 3418–3420.
3. Stephen, A. M. and Cummings, J. H. (1980). The microbial contribution to human faecal mass. *Journal of Medical Microbiology*, 13(1), 45–56.
4. Brown, N. J. (1997). *Health Hazard Manual: Wastewater Treatment Plant and Sewer Workers*. Ithaca, NY: Cornell University, Chemical Hazard Information Program.
5. Cliver, D. O. (1980). Infection with minimal quantities of pathogens from wastewater aerosols. In: *Proceedings of a Symposium on Wastewater Aerosols and Disease*, Cincinnati, OH, September 19–21, 1979, eds. H. Pahren and W. Jakubowski. EPA-600/9-80-028. Environmental Protection Agency, Health Effects Research Laboratory: Cincinnati, OH.
6. Khuder, S. A., Arthur, T., Bisesi, M. S., and Schaub, E. A. (1998). Prevalence of infectious diseases and associated symptoms in wastewater treatment workers. *American Journal of Industrial Medicine*, 33(6), 571–577.
7. Health and Safety Executive (HSE). (2011). Working with sewage. Publication INDG198. Health and Safety Executive: Liverpool, UK.
8. Phair, J. P. (1980). Infection and resistance: A review. In: *Proceedings of a Symposium on Wastewater Aerosols and Disease*, Cincinnati, OH, September 19–21, 1979, eds. H. Pahren and W. Jakubowski. EPA-600/9-80-028. Environmental Protection Agency, Health Effects Research Laboratory: Cincinnati, OH.
9. Sorber, C. A. and Sagik, B. P. (1980). Indicators and pathogens in wastewater aerosols and factors affecting survivability. In: *Proceedings of a Symposium on Wastewater Aerosols and Disease*, Cincinnati, OH, September 19–21, 1979, eds. H. Pahren and W. Jakubowski. EPA-600/9-80-028. Environmental Protection Agency, Health Effects Research Laboratory: Cincinnati, OH.

10. The Center to Protect Workers' Rights (CPWR). (2004). Biological hazards in sewage and wastewater treatment plants, publ. Sewage-9/22/04. The Center to Protect Workers' Rights (CPWR), Building and Construction Trades Department, AFL-CIO: Silver Spring, MD.

11. De Serres, G., Levesque, B., Higgins, R., Major, M., Laliberte, D., Boulianne, N., and Duval, B. (1995). Need for vaccination of sewer workers against leptospirosis and hepatitis A. *Occupational and Environmental Medicine*, 52(8), 505–507.

12. Akin, E. W., Jakubowski, W., Lucas, J. B., and Pahren, H. R. (1978). Health hazards associated with wastewater effluents and sludge: Microbial considerations, in *Risk Assessment and Health Effects of Land Application of Municipal Wastewater and Sludges*, eds. B. P. Sagik and C. A. Sorber. Center for Applied Research and Technology, The University of Texas at San Antonio: San Antonio, TX, pp. 9–25.

13. Safety, Health, and Security in Wastewater Systems Task Force of the Water Environment Federation (WEF). (2013). *Safety, Health, and Security in Wastewater Systems: WEF Manual of Practice No. 1*, 6th ed. McGraw-Hill Education: New York.

14. Clark, C. S. (1987). Potential and actual biological related health risks of wastewater industry employment. *Journal of the Water Pollution Control Federation*, 59, 999–1008.

15. West, P. A., and Locke, R. (1990). Occupational risks from infectious diseases in the water industry. *Water and Environment Journal*, 4(6), 520–523.

16. Jeggli, S., Steiner, D., Joller, H., Tschopp, A., Steffen, R., and Hotz, P. (2004). Hepatitis E, *Helicobacter pylori*, and gastrointestinal symptoms in workers exposed to waste water. *Occupational and Environmental Medicine*, 61(7), 622–627.

17. Niedringhaus, L. and Niedringhaus, L. (1986). Biological hazards at treatment plants. *Operations Forum*, 3(2), 16.

18. Rehman, R. U., Akhtar, N., Akram, M., Shah, P. A., Saeed, T., Jabeen, Q., Asif, H. M., and Shah, S. M. A. (2011). Bacillary dysentery: A review. *Journal of Medicinal Plants Research*, 5(19), 4704–4708.

19. Friis, L., Engstrand, L., and Edling, C. (1996). Prevalence of Helicobacter pylori infection among sewage workers. *Scandinavian Journal of Work, Environment & Health*, 22, 364–368.

20. Friis, L., Agreus, L., and Edling, C. (1998). Abdominal symptoms among sewage workers. *Occupational Medicine*, 48(4), 251–253.

21. Lundholm, M. and Rylander, R. (1983). Work related symptoms among sewage workers. *British Journal of Industrial Medicine*, 40(3), 325–329.

22. Mattsby, I. and Rylander, R. (1978). Clinical and immunological findings in workers exposed to sewage dust. *Journal of Occupational Medicine*, 20, 690–692.

23. Thorn, J., Beijer, L., Jonsson, T., and Rylander, R. (2002). Measurement strategies for the determination of airborne bacterial endotoxin in sewage treatment plants. *Annals of Occupational Hygiene*, 46(6), 549–554.

24. Gerardi, M. H., Maczuga, A. P., and Zimmerman, M. C. (1988). Operator's guide to wastewater viruses. *Public Works PUWOAH*, 119(4), 50.

25. Slade, J. S., Pike, E. B., Eglin, R. P., Colbourne, J. S., and Kurtz, J. B. (1989). The survival of human immunodeficiency virus in water, sewage and sea water. *Water Science and Technology*, 21(3), 55–59.

26. Uhrbrand, K., Schultz, A. C., and Madsen, A. M. (2011). Exposure to airborne noroviruses and other bioaerosol components at a wastewater treatment plant in Denmark. *Food and Environmental Virology*, 3(3–4), 130–137.

27. Centers for Disease Control and Prevention (CDC). (2016). CDC A–Z index: Hand-foot-mouth. Centers for Disease Control and Prevention (CDC), Department of Health and Human Services: Washington, DC. Available: http://www.cdc.gov/hand-foot-mouth/about/index.html. Accessed February 05, 2016.

28. Divizia, M., Cencioni, B., Palombi, L., and Panà, A. (2008). Sewage workers: Risk of acquiring enteric virus infections including hepatitis A. *The New Microbiologica*, 31(3), 337.

29. Santos, L., Santos-Martínez, G., Magaña-Ortíz, J. E., and Puente-Piñón, S. L. (2013). Acute histoplasmosis in three Mexican sewer workers. *Occupational Medicine*, 63(1), 77–79.

30. Centers for Disease Control and Prevention (CDC). (2016). CDC A–Z index: Histoplasmosis. Centers for Disease Control and Prevention, Department of Health and Human Services: Washington, DC. Available: http://www.cdc.gov/fungal/diseases/histoplasmosis/statistics.html. Accessed February 07, 2016.

31. Manos, N. E., Ferebee, S. H., and Kerschbaum, W. F. (1956). Geographic variation in the prevalence of histoplasmin sensitivity. *CHEST Journal*, 29(6), 649–668.

32. Taylor, R. L. (1962). Geographic variation in the prevalence of histoplasmin sensitivity in the Panama Canal Zone. *American Journal of Tropical Medicine and Hygiene*, 11(5), 670–675.

33. Scallan, E., Hoekstra, R. M., Angulo, F. J., Tauxe, R. V., Widdowson, M. A., Roy, S. L., Jones, J. L., and Griffin, P. M. (2011). Foodborne illness acquired in the United States—Major pathogens. *Emerging Infectious Diseases*, 17(1), 7–15.

34. Pecson, B. M., Barrios, J. A., Jiménez, B. E., and Nelson, K. L. (2007). The effects of temperature, pH, and ammonia concentration on the inactivation of Ascaris eggs in sewage sludge. *Water Research*, 41(13), 2893–2902.

35. Clark, C. S., Linnemann, Jr., C. C., Clark, J. G., and Gartside, P. S. (1984). Enteric parasites in workers occupationally exposed to sewage. *Journal of Occupational and Environmental Medicine*, 26(4), 273–275.

36. Clark, C. S., Bjornson, A. B., Schiff, G. M., Pair, J. P., Van-Meer, G. L., and Gartside, P. S. (1977). Sewage worker's syndrome. *Lancet*, 1, 1009.

37. Clark, C. S., Van Meer, G. L., Linnemann, Jr., C. C., Bjornson, A. B., Gartside, P. S., Schiff, G. M., Trimble, S. E., Alexander, D., Cleary, E. J., and Phair, J. P. (1980) Health effects of occupational exposure to wastewater. In: *Proceedings of a Symposium on Wastewater Aerosols and Disease*, Cincinnati, OH, September 19–21, 1979, eds. H. Pahren and W. Jakubowski. EPA-600/9-80-028. Environmental Protection Agency, Health Effects Research Laboratory: Cincinnati, OH.

38. Clark, C. S., Linnemann, Jr., C. C., Gartside, P. S., Phair, J. P., Blacklow, N., and Zeiss, C. R. (1985). Serologic survey of rotavirus, Norwalk agent and *Prototheca wickerhamii* in wastewater workers. *American Journal of Public Health*, 75(1), 83–85.

39. Scarlett-Kranz, J. M., Babish, J. G., Strickl, D., and Lisk, D. J. (1987). Health among municipal sewage and water treatment workers. *Toxicology and Industrial Health*, 3(3), 311–319.

40. Nethercott, J. R. and Holness, D. L. (1988). Health status of a group of sewage treatment workers in Toronto, Canada. *The American Industrial Hygiene Association Journal*, 49(7), 346–350.

41. Zuskin, E., Mustajbegovic, J., and Schachter, E. N. (1993). Respiratory function in sewage workers. *American Journal of Industrial Medicine*, 23(5), 751–761.

42. Douwes, J., Mannetje, A. T., and Heederik, D. (2001). Work-related symptoms in sewage treatment workers. *Annals of Agricultural and Environmental Medicine*, 8(1), 39–45.

43. Heldal, K. K., Madso, L., Huser, P. O., and Eduard, W. (2010). Exposure, symptoms and airway inflammation among sewage workers. *Annals of Agricultural and Environmental Medicine*, 17(2), 263–268.

44. Thorn, J. and Kerekes, E. (2001). Health effects among employees in sewage treatment plants: A literature survey. *American Journal of Industrial Medicine*, 40(2), 170–179.

45. DECOS. (1998). Endotoxins: Health based recommended exposure limit. A report of the Health Council of the Netherlands. Publication No. 1998/03WGD. Health Council of the Netherlands: Rijswijk, the Netherlands.

11

Viral Hepatitis

Hepatitis is a broad term for inflammation of the liver. For this book, we use the word "hepatitis" to denote only those forms caused by a virus.

11.1 Introduction

The disease hepatitis is characterized by inflammation of the liver. The disease causes considerable mortality across the globe, both as acute infection and due to its chronic sequelae [1].

There are six unrelated types of viral hepatitis: A, B, C, D, E, and G. Table 11.1 gives a brief overview.

11.2 Hepatitis A ("Infectious Hepatitis," HAV)

Hepatitis A is an acute, short-term infection that does not lead to long-term liver disease.

Hepatitis A virus (HAV) enters the body via ingestion and spreads to its primary target organ, the liver. There, it incubates for about four weeks. During this time, the virus replicates in the liver and large quantities are shed in the feces. Excretion of virus in the feces declines around the time that symptoms start. Once symptoms start, they usually last one to three weeks; severity ranges from asymptomatic to icteric hepatitis.

Once you have had hepatitis A, you have a certain immunity from reinfection to hepatitis A.

11.2.1 Do Sewage Workers Face an Elevated Risk from HAV?

Are sewage workers at a higher risk from HAV than the general population? The question has been much studied since active hepatitis A vaccines became available [3]. If there is a risk of occupational HAV transmission, then the case for vaccination is clear. However, if sewage workers do not run a

TABLE 11.1

Overview of the Six Unrelated Types of Viral Hepatitis

Type of Viral Hepatitis	Agent	Transmission	Chronic Infection	Vaccine
A	HAV	*Fecal–oral*: Ingestion of water or food contaminated by feces from infected persons	No	Yes
B	HBV	*Blood-borne*: Infected blood, needles; sexual contact; infected mother to newborn	Yes	Yes
C	HCV	*Blood-borne*: Infected blood, needles; sexual contact (much less common mode of transmission) [2]; infected mother to newborn (much less common mode of transmission) [2]	Yes	No
D	HDV	*Blood-borne*: Infected blood, needles; sexual contact; infected mother to newborn		No
E	HEV	*Fecal–oral*: Ingestion of water or food contaminated by feces from infected persons	No	No[a]
G	GBV	*Blood-borne*: Infected blood, needles; sexual contact; infected mother to newborn	Yes	No

[a] In 2011 a vaccine was registered in China [2].

higher risk of occupational transmission than the general population, then why subject them to the possible side effects of vaccination?

The results of several serologic surveys conducted among wastewater employees and control groups are briefly summarized in Table 11.2. The data is mixed. Some studies found that employees who had been exposed to sewage had an elevated risk for HAV infection [4–6]. Others found no increase in the prevalence of anti-HAV among the employees [3,7,8].

11.2.1.1 Are We Comparing Apples and Oranges?

It would be good to have more data about the groups designated "sewage worker" in each study. For example, people working with dewatering sludge should have significantly more biohazard exposure than those who operate the chemical metering equipment. Are they bracketed together?

The study by Venczel et al. [15] in Table 11.2 illustrates the problem. They compare the anti-HAV prevalence among 365 wastewater workers to that of a control group of 166 road and drainage workers. Anti-HAV was found in 38% of the wastewater workers and in 35% of the road workers—apparently no difference. But when they divided wastewater workers into treatment plant workers or sewer workers, the sewer workers showed higher prevalence of anti-HAV, 45%.

TABLE 11.2

Results of Serologic Surveys Conducted among Wastewater Employees and Control Groups

Location	Comparison	Hepatitis Studied	Results	Reference
England	40 sewage workers vs control group of 18 road workers	A	Anti-HAV more prevalent among sewage workers (58%) than control group (33%). Authors conclude that exposure to sewage is a risk factor for hepatitis A infection.	[4]
Singapore	600 sewage workers vs control group of 453 other workers	A	Anti-HAV more prevalent among sewage workers (73%) than control group (51%). Significant factors positively associated with anti-HAV among wastewater workers included >10 years in the wastewater industry and level of education.	[5]
Quebec City, Canada	76 sewage workers vs control group of 152 other workers	A	Anti-HAV was not more prevalent among sewage workers (54%) than among control group (49%).[a] Leptospirosis antibodies were significantly more prevalent among sewage workers (12%) than control group (2%). Age was a significant factor positively associated with anti-HAV among both groups. Authors note a generalized decrease in HAV incidence in Quebec population over the preceding three decades. (*Leptospira* still seems to be abundant.)	[9]
Paris, France	155 sewage workers vs control group of 70 other workers	A	Found that exposure to sewage is an independent risk factor for HAV seropositivity: adjusted odds ratio for HAV seropositivity was 2.15 times greater in sewage workers than control group. Anti-HAV prevalence among sewage workers was 12.9% higher than in control group. This just missed being statistically significant ($p = 0.07$).	[11]

(Continued)

TABLE 11.2 (Continued)

Results of Serologic Surveys Conducted among Wastewater Employees and Control Groups

Location	Comparison	Hepatitis Studied	Results	Reference
England	157 sewage workers vs control group of drinking water workers	A	Significant factors positively associated with anti-HAV among sewage workers include years in the industry, number of siblings, and spending at least 3 months in a country where HAV is endemic. Workers exposed to raw sewage had significantly increased risk of hepatitis A infection. Workers exposed to treated sewage had no increased risk of hepatitis A infection.	[12]
Israel	All 1993–1994 cases reported to Ministry of Health[b] were sorted by occupation	A	Sewage workers did not show any significantly increased risk.	[3]
United States	302 participants; wastewater workers vs control group	A	Wastewater work not significantly associated with antibodies to HAV (anti-HAV), compared to control group. Within group "wastewater workers" no statistically significant occupational risk factors for anti-HAV were identified.	[7]
Texas, United States	359 wastewater workers vs 89 drinking water workers	A	Anti-HAV was more prevalent among wastewater workers than drinking water workers. Seropositivity among wastewater workers was significantly associated with the following: never eating in a lunchroom, ≥8 years in the wastewater industry, never wearing face protection, and skin contact with sewage at least once per day.	[6]

(Continued)

TABLE 11.2 (*Continued*)

Results of Serologic Surveys Conducted among Wastewater Employees and Control Groups

Location	Comparison	Hepatitis Studied	Results	Reference
Israel	100 sewage workers vs control group of 100 blue-collar workers	A	Hepatitis A seropositivity was highly prevalent in both groups. (Israel is an endemic area for the disease.) Among sewage workers, seropositivity was associated with increasing age. Authors conclude that exposure to sewage is not a risk factor for hepatitis A infection in Israel.	[8]
Tuscany, Italy	65 sewage workers vs control group of 160 other workers	A	Difference in anti-HAV prevalence not statistically significant: 51% in sewage workers, 44% in control group. Seropositivity was associated with increasing age, lower education, and birth in southern Italy.[c] Authors note that Italy has had a higher degree of viral circulation in the past. They recommend maintaining plans for HAV immunization of sewage workers, due to changing epidemiology in Italy.	[13]
Georgia, United States	365 sewage workers vs control group of 166 road and drainage workers	A	Anti-HAV was found in 38% of the wastewater workers and in 35% of the road workers. When they divided wastewater workers into treatment plant workers or sewer workers, the sewer workers showed higher prevalence of anti-HAV, 45%.	[15]
Northeastern Italy	66 sewage workers vs control group of 72 other workers	A	Difference in anti-HAV prevalence not statistically significant: 38% in sewage workers, 36% in control group. Seropositivity was associated with increasing age. *Coxsackievirus B3* and *Coxsackievirus B2* were significantly associated with sewage work.	[14]

(*Continued*)

TABLE 11.2 (*Continued*)

Results of Serologic Surveys Conducted among Wastewater Employees and Control Groups

Location	Comparison	Hepatitis Studied	Results	Reference
Copenhagen, Denmark	77 sewer workers vs 81 gardeners vs 79 clerks	A	Anti-HAV significantly higher among sewer workers (80%) than gardeners (60%) or clerks (48%). Seropositivity correlated with age, not years of employment.	[16]
		B	Hepatitis B serological markers were similar in each group. Authors conclude hepatitis B is not successfully transmitted by this route.	[16]
India	92 sewage workers vs control of 55 other workers (no PPE used)	E	Anti-HEV prevalence was significantly higher among sewage workers (56.5%) than among controls (19%). Significant factors positively associated with anti-HAV among wastewater workers: >5 years in the wastewater industry, skin contact with sewage.	[17]
Switzerland	349 sewage workers vs 429 other workers	E	Anti-HEV prevalence was not significantly higher among sewage workers than among controls.	[18]

[a] This study was conducted in 1993. The hepatitis A vaccine was introduced into Canada in 1994. Neither workers nor controls had been vaccinated [10].
[b] Hepatitis A is a notifiable disease in Israel [3].
[c] Hepatitis A is endemic in the Mediterranean area [14].

And what of the waste itself, is it treated or raw human waste? Brugha et al. [12] found that workers who were exposed to raw sewage had significantly increased risk of hepatitis A infection. However, in the same study, workers who were exposed to treated sewage had no increased risk of hepatitis A infection.

It would also be good to know the protective work practices at the sites involved in these studies. It seems quite feasible that if there is a good system of protective work practices at a company, then no elevated risk would be found among their employees.

11.2.1.2 What Conclusions Can We Draw?

It seems reasonable to postulate that if a lot of people in the community have hepatitis A, then the risk may be higher than usual, and the risk will first be seen in sewers. This is because of the following:

- HAV is passed in the feces in large quantities, before the infected person feels ill [1].
- HAV can be stable in the environment for at least 30 days, depending on the environmental conditions [19].
- Wastewater personnel have daily contact with wastewater and therefore a higher potential incidence of exposure than the general population [20].

An example of this is reported by De Serres and Laliberté in their report of a small community outbreak in Quebec City [21]. In this case, 16 people contracted hepatitis B from a contaminated well. Three sewage workers, who had no contact with the original outbreak and belonged to no risk groups, also became ill with the disease. After a thorough investigation eliminating other sources, the authors conclude that despite a normally low incidence of HAV in the community, HAV must be regarded as an occupational hazard for sewage workers, because this group is at elevated risk during an outbreak.

11.2.2 Symptoms

Symptoms of hepatitis A include aches and pains, fever, nausea, loss of appetite, abdominal pain, and jaundice (yellowing of the eyes and skin) [20,22,23]. The severity of infection varies widely, from subclinical or mild illness in children to the full range of symptoms with jaundice in adults [1]. Patients may be incapacitated for several, even many, weeks.

There is no evidence that hepatitis A progresses to chronic liver damage [1].

11.2.3 Prevention

Vaccination and good hygiene.

11.3 Hepatitis B (HBV)

Hepatitis B (HBV) is a blood-borne disease that is endemic in the human population and hyperendemic in many parts of the world. It causes acute and chronic infection of the liver [1].

Hepatitis B can range from a mild illness lasting a few weeks to a serious, lifelong illness [24]. Sequelae can include chronic active hepatitis and cirrhosis. Hepatitis B virus (HBV) is closely associated with hepatocellular carcinoma [1]. The outcome (acute or chronic) seems to depend upon the age at which it is contracted [23]:

- In 95% of adults who get hepatitis B, the immune system will successfully clear the body of the virus ("acute"). Their symptoms will clear up and they will not develop chronic hepatitis B or pass it to others. They will be immune to hepatitis B in the future.
- In 90% of newborns who have hepatitis B, the immune system is too immature; they develop chronic hepatitis B.

HBV is a blood-borne virus and is usually spread when blood, semen, or another body fluid from a person infected with the virus enters the body of someone else.

HBV can survive for at least seven days outside the human body [25].

11.3.1 Symptoms

Symptoms of hepatitis B can include general aches and pains, fever, nausea, loss of appetite, abdominal discomfort, jaundice, and light-colored feces and dark urine [22,23].

11.3.2 Prevention

Vaccination is the best way to prevent hepatitis B [24].

In general, the disease has not been linked to exposure to sewage in the United States [22].

11.4 Hepatitis C

Hepatitis C is a blood-borne virus; blood that contains Hepatitis C virus (HCV) must directly enter the bloodstream of another person to infect them [23]. Sequelae can include chronic active hepatitis and cirrhosis. It is associated with hepatocellular carcinoma in some parts of world [1].

HCV is not one virus; instead it is a complex family of distinct but highly related genotypes and subtypes. The differences in HCV groups could underpin the differences in pathogenicity and in response to antiviral therapy [1]. The virus genotype diversity causes problems in developing a vaccine. Neutralizing antibodies have not been identified so far [1].

11.4.1 Symptoms

Symptoms of hepatitis C include general aches and pains, yellowing of the eyes and skin, nausea, loss of appetite, abdominal discomfort, and light-colored feces and dark urine [23].

11.4.2 Prevention

Avoid blood-to-blood contact [23]. There is no vaccine for hepatitis C. Unlike hepatitis A and hepatitis B, having hepatitis C does not confer immunity to contracting it again.

11.5 Hepatitis D

Hepatitis D is a serious liver disease; sequelae can include chronic active hepatitis and cirrhosis [1].

Although it is caused by the hepatitis delta virus (HDV), it relies on the HBV to replicate. Therefore, you do not contract hepatitis D unless you have hepatitis B.

Hepatitis D is transmitted in a similar fashion to hepatitis B, that is, contact with infected blood.

11.5.1 Two Forms

There are two forms of delta hepatitis infection [1]:

1. Coinfection by HBV and HDV. Individuals who contract both at the same time seem to suffer a more severe form of acute HBV-caused hepatitis. The vaccine for hepatitis B will protect against coinfection.
2. Individuals who suffer chronic infections of HBV may become "superinfected" with HDV. This seems to amplify the hepatitis B: it can cause a second round of clinical hepatitis and accelerate the course of the chronic liver disease or cause overt disease in HBV carriers who had previously been asymptomatic.

11.5.2 Prevention

Avoid contact with infected blood. There is no vaccine for hepatitis D.

11.6 Hepatitis E

Hepatitis E is a fairly new disease, first cloned in 1991 [1]. It is a serious liver disease caused by the hepatitis E virus (HEV), usually with an acute infection, rather than chronic infections.

HEV is rare in North America (though outbreaks have been reported in Mexico), but HEV causes epidemics in the Indian subcontinent, the Middle East, Central and Southeast Asia, and parts of Africa [1].

Having hepatitis E does not seem to confer immunity to contracting it again [26].

11.6.1 HEV and Pregnancy

HEV has a high mortality if contracted during pregnancy, especially during the third trimester [1,2,18].

11.6.2 Prevention

Good hygiene. Currently, there is no FDA-approved HEV vaccine, but progress is being made in this field [1].

11.7 Hepatitis G

Hepatitis G or GB hepatitis is a relatively recent discovery. The virus was first seen in the 1960s in a young surgeon whose initials were "GB." However, it was not until very recently, when DNA techniques were available, that it was demonstrated that these viruses are not genotypes of the HCV.

There are a number of variants: GBV-A, GBV-B, and GBV-C. GBV-A and GBV-C are closely related and sometimes referred to as "GBV-A/C" [1]. Some references exist in the literature to "HGV." GBV-C and HGV are independent isolates of the same virus and are sometimes referred to as "GBV-C/HGV" [27].

Infection can be acute or chronic. A direct association with liver pathology is still lacking [27,28].

Important work that still needs to be done with this new virus include [1,27]

- Development of specific diagnostic reagents
- Standardizing detection methods
- Establishing the epidemiology

- Determining the pathogenic significance in humans of GBV
- Clarifying the role of this virus in coinfections with other viruses
- Determining the clinical and public health importance of GBV

References

1. Zuckerman, A. J. (1996). Hepatitis viruses, Chapter 70, in *Medical Microbiology*, 4th ed., ed. S. Baron. University of Texas Medical Branch at Galveston: Galveston, TX. Available: http://www.ncbi.nlm.nih.gov/books/NBK7864/. Accessed February 08, 2016.
2. WHO. (2015). Hepatitis E. Fact Sheet No. 280. World Health Organization: Geneva, Switzerland. Available: www.who.int/mediacentre/factsheets/fs280/en. Accessed February 03, 2016.
3. Lerman, Y., Chodik, G., Aloni, H., Ribak, J., and Ashkenazi, S. (1999). Occupations at increased risk of hepatitis A: A 2-year nationwide historical prospective study. *American Journal of Epidemiology*, 150(3), 312–320.
4. Poole, C. J. and Shakespeare, A. T. (1993). Should sewage workers and carers for people with learning disabilities be vaccinated for hepatitis A? *British Medical Journal*, 306(6885), 1102.
5. Heng, B. H., Goh, K. T., Doraisingham, S., and Quek, G. H. (1994). Prevalence of hepatitis A virus infection among sewage workers in Singapore. *Epidemiology and Infection*, 113(1), 121–128.
6. Weldon, M., VanEgdom, M. J., Hendricks, K. A., Regner, G., Bell, B. P., and Sehulster, L. M. (2000). Prevalence of antibody to hepatitis A virus in drinking water workers and wastewater workers in Texas from 1996 to 1997. *Journal of Occupational and Environmental Medicine*, 42(8), 821–826.
7. Trout, D., Mueller, C., Venczel, L., and Krake, A. (2000). Evaluation of occupational transmission of hepatitis A virus among wastewater workers. *Journal of Occupational and Environmental Medicine*, 42(1), 83.
8. Levin, M., Froom, P., Trajber, I., Lahat, N., Askenazi, S., and Lerman, Y. (2000). Risk of hepatitis A virus infection among sewage workers in Israel. *Archives of Environmental Health: An International Journal*, 55(1), 7–10.
9. De Serres, G., Levesque, B., Higgins, R., Major, M., Laliberte, D., Boulianne, N., and Duval, B. (1995). Need for vaccination of sewer workers against leptospirosis and hepatitis A. *Occupational and Environmental Medicine*, 52(8), 505–507.
10. Nicol, A. G., Wright, M. E., Prentice, A. C., Carroll, A., Kemp, S., and Reed, J. M. (1996). [Correspondence] Need for vaccination of sewer workers against leptospirosis and hepatitis A. *Occupational and Environmental Medicine*, 53(1), 71.
11. Cadilhac, P. and Roudot-Thoraval, F. (1996). Seroprevalence of hepatitis A virus infection among sewage workers in the Parisian area, France. *European Journal of Epidemiology*, 12(3), 237–240.
12. Brugha, R., Heptonstall, J., Farrington, P., Andren, S., Perry, K., and Parry, J. (1998). Risk of hepatitis A infection in sewage workers. *Occupational and Environmental Medicine*, 55(8), 567–569.

13. Bonanni, P., Comodo, N., Pasqui, R., Vassalle, U., Farina, G., Nostro, A. L., Boddi, V., and Tiscione, E. (2000). Prevalence of hepatitis A virus infection in sewage plant workers of Central Italy: Is indication for vaccination justified? *Vaccine*, 19(7), 844–849.

14. Divizia, M., Cencioni, B., Palombi, L., and Panà, A. (2008). Sewage workers: Risk of acquiring enteric virus infections including hepatitis A. *The New Microbiologica*, 31(3), 337.

15. Venczel, L., Brown, S., Frumkin, H., Simmonds-Diaz, J., Deitchman, S., and Bell, B. P. (2003). Prevalence of hepatitis A virus infection among sewage workers in Georgia. *American Journal of Industrial Medicine*, 43(2), 172–178.

16. Skinhøj, P., Hollinger, F. B., Hovind-Hougen, K., and Lous, P. (1981). Infectious liver diseases in three groups of Copenhagen workers: Correlation of hepatitis A infection to sewage exposure. *Archives of Environmental Health: An International Journal*, 36(3), 139–143.

17. Vaidya, S. R., Tilekar, B. N., Walimbe, A. M., and Arankalle, V. A. (2003). Increased risk of hepatitis E in sewage workers from India. *Journal of Occupational and Environmental Medicine*, 45(11), 1167–1170.

18. Jeggli, S., Steiner, D., Joller, H., Tschopp, A., Steffen, R., and Hotz, P. (2004). Hepatitis E, *Helicobacter pylori*, and gastrointestinal symptoms in workers exposed to waste water. *Occupational and Environmental Medicine*, 61(7), 622–627.

19. McCaustland, K. A., Bond, W. W., Bradley, D. W., Ebert, J. W., and Maynard, J. E. (1982). Survival of hepatitis A virus in feces after drying and storage for 1 month. *Journal of Clinical Microbiology*, 16(5), 957–958.

20. WEF. Safety, Health, and Security in Wastewater Systems Task Force of the Water Environment Federation (2013). *Safety, Health, and Security in Wastewater Systems: WEF Manual of Practice No. 1*, 6th ed. McGraw-Hill Education: New York.

21. De Serres, G. and Laliberté, D. (1997). Hepatitis A among workers from a waste water treatment plant during a small community outbreak. *Occupational and Environmental Medicine*, 54(1), 60–62.

22. The Center to Protect Workers' Rights (CPWR). (2004). Biological hazards in sewage and wastewater treatment plants, publ. Sewage-9/22/04. The Center to Protect Workers' Rights (CPWR), Building and Construction Trades Department, AFL-CIO: Silver Spring, MD.

23. Hepatitis Queensland. (2016). Hepatitis C (HCV). South Brisbane, Queensland, Australia. Available: http://hepqld.asn.au/component/content/article/6-hepatitis. Accessed February 01, 2016.

24. National Institute for Occupational Safety and Health (NIOSH). (2011). Health hazard evaluation report: Evaluation of exposures associated with cleaning and maintaining composting toilets—Arizona. NIOSH HETA No. 2009-0100-3135. By Burton, N. C. and Dowell, C. U.S. Department of Health and Human Services, Centers for Disease Control and Prevention, National Institute for Occupational Safety and Health: Cincinnati, OH.

25. Bond, W. W., Favero, M. S., Petersen, N. J., Gravelle, C. R., Ebert, J. W., and Maynard, J. E. (1981). Survival of hepatitis B virus after drying and storage for one week. *Lancet*, 1(8219), 550–551.

26. Jameel, S., Durgapal, H., Habibullah, C. M., Khuroo, M. S., and Panda, S. K. (1992). Enteric non-A, non-B hepatitis: Epidemics, animal transmission, and hepatitis E virus detection by the polymerase chain reaction. *Journal of Medical Virology*, 37(4), 263–270.
27. Sathar, M. A., Soni, P. N., and York, D. (2000). GB virus C/hepatitis G virus (GBV-C/HGV): Still looking for a disease. *International Journal of Experimental Pathology*, 81(5), 305–322.
28. Reshetnyak, V. I., Karlovich, T. I., and Ilchenko, L. U. (2008). Hepatitis G virus. *World Journal of Gastroenterology*, 14(30), 4725–4734.

12

Leptospirosis/Weill's Disease

12.1 Introduction

The disease leptospirosis is a zoonosis: it is always acquired from an animal source. Man can be an incidental host [1,2] but generally human-to-human transmission can be considered nonexistent [3,4]. Humans almost never function as carriers, but suffer acute infection, sometimes with long-term sequelae [4].

It is a systemic disease characterized by fever, renal and liver damage, pulmonary hemorrhage, and reproductive failure. The illness may be relatively mild, resembling influenza, or in its severe form (Weil's disease) may lead to fatal pulmonary hemorrhage or renal or liver failure. The severity seems to depend on the infecting serovar* of *Leptospira*, the immune system it encounters, and the health and age of the patient.

12.2 The Pathogen and How It Spreads

Leptospirosis is caused by leptospire-type bacteria: long thin spirochetes that are members of the genus *Leptospira*. There are at least 12 pathogenic species [4]. Leptospires have an optimum growth temperature of 28°C–30°C and prefer neutral pH [4,5].

Leptospires thrive in the kidneys of carrier animals. The most important host animals, from a human perspective, are rats; dogs, pigs, and cattle are also carriers. The leptospires can also live freely outside of host animals, as long as they are in moist conditions, such as surface waters, mud, or sewers.

---

* Variations within a species.

12.2.1 Carrier Animals

Carriers may be wild or domestic animals: the most important are rodents, small marsupials, cattle, pigs, and dogs [4]. In urban environments, the major potential reservoir mammals are rats and dogs [3].

For humans, the most important carriers are mice and rats, mainly *Rattus norvegicus* and *Rattus rattus*, because they serve as reservoirs for spreading the disease and they tend to carry virulent strains. They are usually symptom-free but harbor the leptospires in their kidneys. From the carriers' kidneys, the leptospires pass into the urine; when the urine is excreted, the leptospires contaminate the soil or waters in which they land [4,6]. Urinary concentrations of the bacteria may be as high as 10^8–10^{10}/mL [1,4].

Animals recovering from leptospirosis may become symptom-free carriers whose renal tubules are colonized by leptospires for extended periods. An infected animal can remain symptom-free and spread infectious leptospires for its entire lifetime [3,4,7].

Carrier animals distribute the leptospires in their urine. Leptospires prefer a neutral pH. They do not survive well in acid urine but remain viable in alkali urine. Animals whose urine is alkaline are relatively more important as carriers than those that have acidic urine [4]. Human urine, incidentally, is acidic [5]; this may help explain why humans can easily become infected, but are not significant carriers.

12.2.2 Infection Routes

Leptospires can enter the body through abrasions or cuts in the skin, through the conjunctiva of the eyes, or through the mucous membranes lining the mouth and nose.

Typical infection routes include

- Breathing aerosols containing the leptospires
- Taking in water during swimming
- Contact with infected animals' urine while caring for livestock or pets
- Contact with infected animals' tissue, for example, in slaughterhouses
- Handling objects that have been in contact with their urine

Occupations traditionally at risk from leptospirosis include sewer workers, miners, farmers, slaughterhouse workers, fish workers, veterinarians and animal caretakers, and military personnel. The implementation of protective measures (pest control and protective clothing) within the sewage industry has done much to decrease the occupational risk. In the British Isles, for example, sewer workers accounted for 79 of 983 leptospirosis cases during 1933–1948. Some 30 years later, the numbers had fallen: only 4 of 358 leptospirosis cases were sewer workers in 1978–1983 [1].

12.3 Symptoms and Course of the Disease

The bacteria spread through the bloodstream and can grow in any or all tissues. When their numbers in the blood and tissues are large enough, lesions appear. The lesions damage small blood vessels, leading to localized ischemia in critical organs; this in turn can cause renal tubular necrosis, destruction of liver cells, pulmonary damage, jaundice, meningitis, myositis, and placentitis [4,6].

The patient experiences a sudden onslaught of headache, fever (typically to 102°F, 39°C), malaise, myalgia, conjunctival suffusion, and sometimes a rash. Leptospirosis can run its course as a biphasic illness (anicteric form) or fulminant disease (icterohemorrhagic form) [3,4,8].

12.3.1 More Knowledge of the Mechanisms Is Needed

Current understanding of the mechanisms of leptospirosis pathogenesis is limited. Areas that need elucidation include the following basic questions [2,3,6]:

- Does the outcome of an infection (i.e., severe or mild) depend upon the direct pathogen effects or the host immune responses?
- What role do the leptospires' toxins play in determining virulence?
- What is the mechanism of host immunity to *Leptospira*, and what role does host immunity play in *Leptospira* pathogenesis?
- How is naturally acquired immunity mediated?
- Is there a correlation between *Leptospira* species and severity of the disease?

The last question might require amplification. The Icterohemorrhagiae serogroup is often implicated in the more severe cases, but some patients infected with it experience only mild symptoms. Other patients, infected with usually mild strains, can suffer severe, even life-threatening, symptoms [2,4]. The infecting strain seems to be a major factor, but not a determinant, of the severity of the disease.

12.3.2 Severe Leptospirosis

Severe leptospirosis is usually associated with serovars Icterohemorrhagiae, Copenhageni, Lai, and others. As mentioned earlier, the infecting strain does not necessarily determine whether the disease will be mild or severe; bewilderingly, a strain that is normally in the "mild" category will occasionally cause life-threatening illness.

After the onset of illness, the patient worsens rapidly. Renal failure may occur within 7–10 days, sometimes together with one or more of the following: liver failure, jaundice, pulmonary hemorrhages, or hemorrhages of the skin or mucosal membranes [3,4,7].

12.3.3 Chronic or Recurrent Leptospirosis

Chronic, long-term sequelae have been reported with both mild and severe leptospirosis. The factors determining long-term persistence, and the mechanism involved, are unknown [4].

12.3.4 Reproductive Failure

Faine et al. [9] have reported that leptospirosis has caused spontaneous abortion. Shaked et al. [10] have reviewed 16 cases of leptospirosis in human pregnancy: three delivered healthy babies, four delivered babies who had signs of active leptospirosis, and eight had spontaneous abortions. (The fate of one fetus was not recorded.) Women appeared to be more likely to spontaneously abort if the leptospirosis occurred in the early months of pregnancy [10].

12.4 Detecting Leptospirosis

Diagnosis of leptospirosis is difficult because of the number of organs that can be attacked (lung, liver, kidney, eye, brain) and the resulting variety of clinical manifestations.

It can mimic the symptoms of many other diseases, for example, dengue, Venezuelan hemorrhagic fever, Mayaro virus disease, psittacosis, parvovirus B19, and Oropouche fever [3,11].

Clinical diagnosis depends upon a variety of laboratory techniques [3–5,12–15]:

- Microscopic agglutination test (MAT): this is the reference standard test because of its high sensitivity and specificity [16].
- Indirect hemagglutination assay (IHA).
- Polymerase chain reaction (PCR) and real-time PCR (qPCR).
- Enzyme-linked immunosorbent assay (ELISA).

Identification by culture is the definitive diagnosis. However, leptospires are slow-growing bacteria and cultures can take up to 13 weeks. This makes identifying the disease by culture impractical for clinical diagnoses [3,4]; it is a useful tool, however, for epidemiologists investigating outbreaks.

12.5 Who Is at Risk?

Humans are at risk if they come into contact with the urine of carrier animals or contaminated water or soil [4]. It is an occupational risk for many: farmers working in moist conditions, attending animals, or milking cows; miners; meat workers; canoeists; and swimmers. Military personnel are often at risk, because areas that suffer civil or military emergencies often have had their rat control systems destroyed and suffer heavy rodent infestation.

12.5.1 Leptospirosis and Water Sports

The disease is also associated with swimming, wading, kayaking, windsurfing, and rafting in contaminated lakes and rivers, both in developed countries and in tropical endemic areas. There have been several reports of outbreaks after athletic events in recent years:

- A large leptospirosis outbreak occurred at the 1998 Springfield Illinois Triathlon [17]. The triathlon involved cycling, swimming (in Lake Springfield), and running. Of the 876 triathletes, 98 reported becoming ill after the competition. Heavy rains preceding the triathlon are thought to have increased leptospiral contamination in Lake Springfield.
- In October 2001, six canoeists contracted leptospirosis after competing in a white-water competition on the River Liffey in Dublin, Ireland [18].
- In November 2005, 14 of 192 athletes contracted leptospirosis participating in an endurance-length swamp race outside of Tampa, Florida. This adventure race involved both swamp water and river water [19].
- In 2006, five athletes contracted the disease after a triathlon in Heidelberg, Germany. The triathlon involved swimming in the Neckar river. Heavy rains had preceded the event [20].
- In July 2010, four athletes contracted the disease after a triathlon in Langau, Austria. Heavy rains had preceded the event [8].
- In 2013, an outbreak of the disease occurred among athletes participating in a triathlon on Réunion Island. Twenty-three percent of the adult swimmers tested positive for leptospirosis. The epidemiological investigation found evidence that complete neoprene suits offered some protection against the disease [14].

The most high-profile case of leptospirosis contracted during water sports is undoubtedly that of Olympic champion rower Andy Holmes. Holmes, who won medals in the Los Angeles and Seoul Olympics, died in October 2010 shortly after his 51st birthday from Weil's disease.

12.6 Preventive Measures

In our industry it is not practicable to advise sewer workers to avoid wading in water that may be contaminated. However, measures for occupational hygiene, such as wearing protective clothing (boots, goggles, overalls, and rubber gloves) and avoiding splashes from water, are often useful.

Vigorous rodent control measures are extremely important in limiting the extent of leptospire contamination.

12.6.1 Vaccines

Vaccines for humans have been in use since the 1920s. The vaccines are not a "magic bullet"—in addition to side effects, the immunity each vaccine can offer is restricted to antigenically related serovars. The geographical distribution of serovars can, and does, change occasionally [2].

A more thorough discussion of the vaccine situation is outside the scope of this chapter, but for the interested reader there is an excellent review by Ko et al. [21].

Repeated annual revaccination is recommended for maintenance of immunity [4].

12.7 Summary

The handling of leptospirosis may be viewed as a success story for our industry, due mainly to the following work practices [22]:

- Energetic rodent control programs, reducing the population of carrier rats
- Vigilant use of protective clothing
- Scrupulous personal hygiene
- General awareness of leptospirosis as an occupational hazard

References

1. Waitkins, S. A. (1985). From the PHLS. Update on leptospirosis. *British Medical Journal*, 290(6480), 1502.
2. Forbes, A. E., Zochowski, W. J., Dubrey, S. W., and Sivaprakasam, V. (2012). Leptospirosis and Weil's disease in the UK. *The Quarterly Journal of Medicine*, 105, 1151–1162.

3. Bharti, A. R. et al. (2003). Leptospirosis: A zoonotic disease of global importance. *The Lancet Infectious Diseases*, 3(12), 757–771.
4. Adler, B. and de la Peña Moctezuma, A. (2010). Leptospira and leptospirosis. *Veterinary Microbiology*, 140(3), 287–296.
5. Waitkins, S. A. (1986). Leptospirosis as an occupational disease. *British Journal of Industrial Medicine*, 43(11), 721.
6. Levett, P. N. and Haake, D. A. (2009). Leptospira species (leptospirosis), in *Mandell, Douglas, and Bennett's Principles and Practice of Infectious Diseases*, 7th ed., eds. Mandell, J.L., Bennett, J.E., and Dolin, R. Churchill Livingstone Elsevier: Philadelphia, PA, pp. 3059–3066.
7. Levett, P. N. (2001). Leptospirosis. *Clinical Microbiology Reviews*, 14, 296–326.
8. Radl, C. et al. (2011). Outbreak of leptospirosis among triathlon participants in Langau, Austria, 2010. *Wiener klinische Wochenschrift*, 123(23–24), 751–755.
9. Faine, S., Adler, B., Christopher, W., and Valentine, R. (1984). Fatal congenital human leptospirosis. *Zentralblatt für Bakteriologie, Mikrobiologie und Hygiene. Series A: Medical Microbiology, Infectious Diseases, Virology, Parasitology*, 257(4), 548.
10. Shaked, Y., Shpilberg, O., Samra, D., and Samra, Y. (1993). Leptospirosis in pregnancy and its effect on the fetus: Case report and review. *Clinical Infectious Diseases*, 17(2), 241–243.
11. Vinetz, J. M. (2010). 10 common questions about leptospirosis. *Infectious Diseases in Clinical Practice*, 9, 19–25.
12. Blanco, R. M., dos Santos, L. F., Galloway, R. L., and Romero, E. C. (2016). Is the microagglutination test (MAT) good for predicting the infecting serogroup for leptospirosis in Brazil? *Comparative Immunology, Microbiology and Infectious Diseases*, 44, 34–36.
13. Kanimozhi, R., Geetha, R., Anitha, D., and Ramesh, V. (2016). A serological study of leptospirosis in Chennai. *International Journal of Research in Medical Sciences*, 4(1), 268–271.
14. Pagès, F. et al. (2016). Investigation of a leptospirosis outbreak in triathlon participants, Réunion Island, 2013. *Epidemiology and Infection*, 144(03), 661–669.
15. Żmudzki, J., Jabłoński, A., Nowak, A., Zębek, S., Arent, Z., Bocian, Ł., and Pejsak, Z. (2016). First overall report of Leptospira infections in wild boars in Poland. *Acta Veterinaria Scandinavica*, 58(1), 1.
16. Weyant, R. S., Bragg, S. L., and Kaufmann, A. F. (1999). Leptospira and Leptonema, in *Manual of Clinical Microbiology*, 7th ed., eds. P. R. Murray, E. J. Baron, M. A. Pfaller, F. C. Tenover, and R. H. Yolken. American Society for Microbiology: Washington, DC, pp. 739–745.
17. Morgan, J. et al. (2002). Outbreak of leptospirosis among triathlon participants and community residents in Springfield, Illinois, 1998. *Clinical Infectious Diseases*, 34(12), 1593–1599.
18. Boland, M., Sayers, G., Coleman, T., Bergin, C., Sheehan, N., Creamer, E., O'Connell, M., Jones, L., and Zochowski, W. (2004). A cluster of leptospirosis cases in canoeists following a competition on the River Liffey. *Epidemiology and Infection*, 132(02), 195–200.
19. Stern, E. J. et al. (2010). Outbreak of leptospirosis among Adventure Race participants in Florida, 2005. *Clinical Infectious Diseases*, 50(6), 843–849.

20. Brockmann, S. et al. (2010). Outbreak of leptospirosis among triathlon participants in Germany, 2006. *BMC Infectious Diseases*, 10(1), 91.
21. Ko, A. I., Goarant, C., and Picardeau, M. (2009). Leptospira: The dawn of the molecular genetics era for an emerging zoonotic pathogen. *Nature Reviews Microbiology*, 7(10), 736–747.
22. West, P. A. and Locke, R. (1990). Occupational risks from infectious diseases in the water industry. *Water and Environment Journal*, 4(6), 520–523.

13

Exercises

13.1 Confined Spaces

1. The manhole shown in Figure 13.1 is considered a potentially hazardous confined space because
   A. Entry and exit are difficult.
   B. The sewer system is not designed for occupation.
   C. The sewer system can contain hazards such as low oxygen or toxic/flammable gases.
   D. All of the above.
   E. It isn't a confined space.

2. In Figure 13.2, the sewer pipe has been lowered to the bottom of a three-meter deep trench but has not yet been connected to the sewer system. A worker has to enter the pipe to inspect a weld. Should it be considered a confined space?
   A. Yes.
   B. Yes. Also beware of excavation hazards!
   C. No. It is not a confined space until connected to the collection system.

3. The primary hazards of collection (sewer) systems are
   A. Toxic or flammable gases
   B. Oxygen deficiency
   C. Machinery
   D. A and B

4. When using a self-contained breathing apparatus (SCBA) in the sewer system, the amount of time you can spend in the collection system depends on
   A. The amount of air in the tanks, and how fast it is consumed when you are in the sewer system.

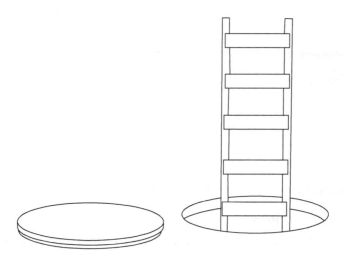

FIGURE 13.1
Collection system manhole.

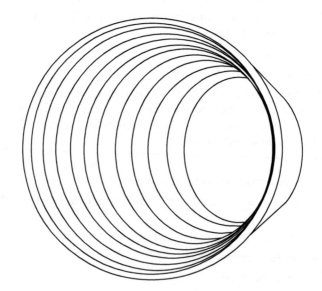

FIGURE 13.2
Sewer pipe.

  B. The task that needs to be done.
  C. Whether the nearest exit point is in front of you or at the entry
     point behind you.
5. Using a supplied-air (air-line) respirator limits
  A. How far you can go in the collection system

B. The amount of tools that you can carry

C. How long you can stay in the collection system

13.1.1 Entering the Confined Space

The questions in this section apply to Figure 13.3.

6. The guardrails are there to protect
   A. The workers shown in the picture, from nearby vehicle traffic
   B. Passersby, who might otherwise fall down the open manhole
   C. The tripod and ladder
   D. All of the above

7. The man standing on the ground holding the rope
   A. Will follow his colleague into the manhole when he gets his gear on
   B. Will stay outside the manhole the whole time
   C. Has done the atmospheric testing and can leave the site as soon as his colleague has safely entered the confined space
   D. Will only enter the confined space if his colleague is in trouble

FIGURE 13.3
Entering the collection system.

8. The man on the ladder is equipped with
   A. An air-purifying cartridge respirator
   B. A supplied-air respirator
   C. A self-contained breathing apparatus
9. The man standing on the ground is not wearing breathing equipment because
   A. It's only needed when you enter the confined space.
   B. He is standing well to the side of the manhole.

13.2 Hydrogen Sulfide

10. The hazards of hydrogen sulfide include
    A. It is toxic.
    B. It is flammable/explosive.
    C. It has a high COD, thereby causing oxygen deficiency.
    D. A and B.
11. The soda-can effect
    A. Is irritating but not really dangerous
    B. Can be lethal because it happens so quickly
    C. Can be lethal because the concentrations of H_2S released from the wastewater can be very high
    D. B and C
12. True or false: If you survive acute H_2S exposure, then you will not suffer any permanent ill effects.
    A. True
    B. False
13. True or false: "Knockdown" is never fatal, nor does it lead to permanent sequelae.
    A. True
    B. False

13.2.1 H₂S Case: Silo for Sludge

This section describes an actual hydrogen sulfide accident reported by Nogué et al. [1]. The case involved a building that contained a silo for sludge from water purification stations.

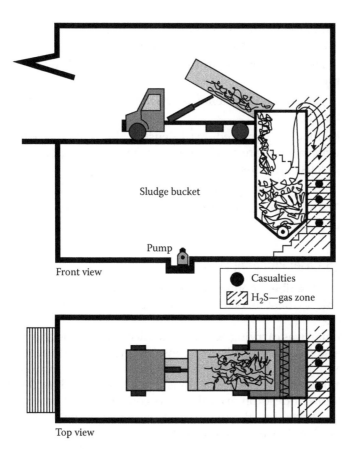

FIGURE 13.4
Truck dumping sludge into silo. (From Nogué, S. et al., *Occup. Med.*, 61(3), 212, 2011.)

The sludge silo had a 120-ton capacity and operated on two stories: on the upper story, trucks drove up to the hopper to unload sludge, and on the lower story, there was the silo outlet and associated equipment such as pumps and valves (see Figure 13.4). The mouth of the hopper was normally closed and opened only for the few minutes when trucks unloaded sludge.

One day, while a worker on the lower level was checking a pump, a truck came in and dumped sludge into the silo. The 27 tons of new sludge fell circa two meters onto the 59 tons of existing sludge. The combination of large volume, height, and speed is believed to have caused H_2S that had accumulated in the silo to rise in a cloud and spill over its sides.

Just as the H_2S cloud spilled over, the worker who had been on the lower level was climbing the stairs to leave the area. The stairs were adjacent to the silo. The worker was overcome by the descending H_2S cloud and lost consciousness on the stairs. Two coworkers rushed to his aid and also lost consciousness. When firefighters arrived, the three workers were dead.

14. Which of the following could have helped prevent this accident? If more than one, list in order of importance.

    A. A gas detection system that warns if H_2S spills over the top of the silo.

    B. A gas detection system that warns if H_2S is building up inside the silo.

    C. Methods and procedures to ensure that staff are not present, when actions are carried out that may generate potentially dangerous emissions.

    D. Training for all operative staff in the hazards of H_2S.

    E. Training for all operative staff in confined spaces and rescue operations.

13.2.2 H₂S Case: Pumping Station

This section describes another hydrogen sulfide accident reported by Nogué et al. [1]. This accident involved a wastewater pumping substation, shown in Figure 13.5.

    The foreman was carrying out a routine check of the substation. As he went down the steps, he suddenly lost consciousness. The emergency services rescued him approx. 15 minutes later, in a coma. After approx. eight hours at a community hospital, he was transferred to another hospital, where he was pronounced dead on arrival. It was concluded that when the substation pumps turned on, H_2S that had accumulated in the wastewater in the underground tank was stirred up, entered the gas phase, and migrated out

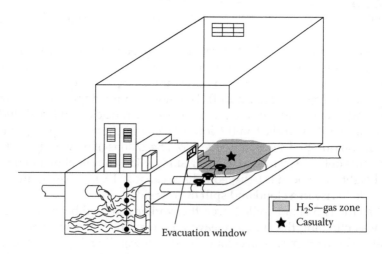

FIGURE 13.5
Pump station layout. (From Nogué, S. et al., *Occup. Med.*, 61(3), 212, 2011.)

of the underground tank via a window into the machinery area of the substation (see Figure 13.5).

15. With the design shown in Figure 13.5, would sufficient protection be provided by gas detection systems that trigger alarms if H_2S enters the machinery area?

   A. Yes

   B. No

16. Is this a confined space accident?

   A. Yes

   B. No

17. Would training in confined spaces have prevented this accident?

   A. Yes

   B. No

18. Which one of the following is most likely to have helped prevent this accident?

   A. A gas detection system connected to powerful ventilation systems (venting safely to the exterior of the building); an alarm from the detector automatically starts the ventilation.

   B. A gas detection system that warns if H_2S is building up in the underground tank.

   C. A pump station design wherein the air in the underground tank can never come into the machinery room.

   D. A ventilation system for the underground tank, where gases extracted would be vented to the exterior of the building in a safe manner.

   E. Training for all operative staff in confined spaces.

13.3 Methane

19. True or false: Methane is only a hazard if it can seep in from the surrounding rock strata.

   A. True

   B. False

20. True or false: Methane is never a concern if the circumstances are always aerobic.

   A. True

   B. False

21. True or false: Methane is not soluble in water.
    A. True
    B. False
22. True or false: If the methane is just below the lower explosive limit (LEL), then it should still be regarded as hazardous.
    A. True
    B. False
23. True or false: If the methane level is well above the upper explosive limit (UEL), then the conditions can be regarded as safe.
    A. True
    B. False

13.4 Carbon Monoxide, Carbon Dioxide

24. The primary danger of operating a diesel engine inside a sewer is
    A. Carbon monoxide poisoning
    B. Carbon dioxide poisoning
    C. Oxygen deficiency
25. Explosive blasting near depressions such as trenches being dug for sewers is dangerous because
    A. The blast uses up the available oxygen in the area, causing oxygen deficiency.
    B. The blasting creates carbon dioxide that is toxic even at very low levels.
    C. The blasting creates carbon monoxide that can migrate through the soil to the sewer trench.
26. Operating a gasoline engine just outside of the sewer manhole can be hazardous because
    A. The amount of noise the engine makes exceeds the decibel limit.
    B. Carbon monoxide produced by the engine can migrate into the manhole and accumulate down in the sewer.
    C. The risk of gasoline spilling and running into the sewer.

13.4.1 CO Case: Gasoline Engine in Sewer

This section describes a fatal carbon monoxide accident reported by the National Institute for Occupational Safety and Health [2] during the construction of a new sewer line. The project involved constructing several thousand feet of 66-inch diameter sanitary sewer and tying it into the existing collection system. When the new sewer was completed, the existing sewer line would be abandoned.

During construction, the existing line had to be kept in service, and the new sewer had to connect to it. In order to facilitate this, a bypass line was tapped into the existing sewer, to divert the flow around the connection point between old and new sewers. The bypass line reentered the collection system at Manhole 1 (see Figure 13.6).

To keep sewage from entering the work area, the pipe was diked by sand-bags several feet upstream of Manhole 1. Sewage seeped past the sandbags, so a steel plug replaced the sandbag dike. To remove the existing sewage, a gasoline-engine-driven pump was brought in and placed upstream of the plug. Maintaining the pump required a worker to enter at Manhole 2, walk approximately 1200 feet to the pump, fuel the gasoline engine, start it, and walk back to the exit at Manhole 2. There was no forced (mechanical) ventilation to remove air contaminants, and at no time was the atmosphere in the pipe tested before entry.

On the day of the accident, a foreman and a worker performed this procedure at 8:30 a.m. When they went to do it again at 3 p.m., Manhole 2 had been covered over with plywood so that concrete could be poured the next day. They entered from the point of construction and began walking 3000 feet to

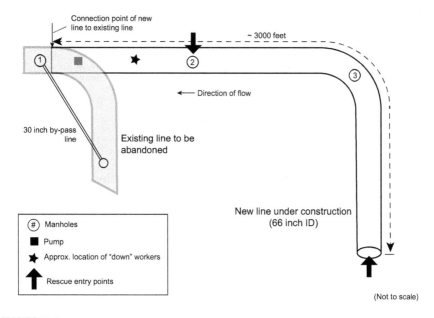

FIGURE 13.6
New and existing sewer lines. Not to scale. (From U.S. Department of Health, Education and Welfare, National Institute for Occupational Safety and Health (NIOSH), Fatal accident summary report: Two confined space fatalities during construction of a sewer line, Fatal accident circumstances and epidemiology database, incident FACE 8413, Washington, DC, 1984, accessed January 14, 2016.)

the pump. At the pump, the worker tried to start the engine four times and then collapsed. The foreman could not carry the worker out, so he propped him up out of the water and walked and crawled 3000 feet to the exit to fetch help. Seven coworkers went into the pipe in unsuccessful attempts to rescue the downed worker. A state inspector went to Manhole 2, removed the plywood, and entered that way in a rescue attempt; he did not come out. The underground superintendent also entered the sewer at Manhole 2 but exited after two to three minutes. The company safety director then entered Manhole 2 and managed to reach the downed worker before he had to turn back.

Three firemen equipped with self-contained breathing apparatus (SCBA) then arrived and entered Manhole 2. The bodies of the state inspector and the downed worker were retrieved. Autopsies showed that both men died of carbon monoxide poisoning.

27. The *primary* cause of the accident was
    A. Failure to prevent sewage from flowing into the newly constructed sewer
    B. Using a gasoline-powered pump inside a sewer, without adequate ventilation
    C. Covering Manhole 2 with plywood too early

28. Instead of placing the gasoline-powered pump inside the sewer, the contractor could have
    A. Used an electric pump
    B. Placed the gasoline-powered pump outside the sewer with a hose running to the sewage
    C. A or B

29. Confined space training could have prevented
    A. The initial entry into a confined space without testing the atmosphere
    B. The death of the worker
    C. Ten people endangering themselves by attempting rescue without SCBA
    D. The death of the state inspector
    E. All of the above

13.5 Biological Hazards

30. True or false: HIV is one of the most important biological hazards in the sewer system.
    A. True
    B. False

31. Important measures to protect yourself against infection by protozoa and worms include
    A. Vaccination
    B. Washing hands before eating, drinking, or smoking
    C. Keeping clean clothes separate from used work clothes
    D. Cleaning personal protective equipment after use
    E. B, C, and D

32. If you have had hepatitis A, then you have immunity against
    A. Hepatitis B
    B. Hepatitis C
    C. Hepatitis E
    D. Hepatitis G
    E. Hepatitis B and D
    F. None of the above

33. Which vaccine will help protect against hepatitis D?
    A. Hepatitis A vaccine
    B. Hepatitis B vaccine
    C. Hepatitis C vaccine
    D. Hepatitis D vaccine
    E. None of the above

34. True or false: Vaccinations against tetanus are not needed in the
wastewater industry.
    A. True
    B. False

35. True or false: *Helicobacter pylori* is a human carcinogen and it has
been proven that it causes peptic ulcer among sewage workers.
    A. True
    B. False

13.6 Answers

1. D. The collection system has all of these features, and more, that can
   make it a potentially hazardous confined space.
2. B. The pipe is a confined space. The trench in which the pipe sits
   is an excavation hazard and precautions against trench collapse
   should be taken.

3. D

4. A. You *must* have enough air left to exit the collection system, whether or not the task is completed.

5. A. You can only go as far as the respirator line.

6. D. Protecting the workers and passersby is the highest priority. It is also necessary to protect the tripod and ladder; without them in position, an emergency evacuation is difficult or even impossible.

7. B

8. C

9. B

10. D. H_2S is highly toxic, flammable, and explosive. It is true that H_2S is created in anaerobic conditions; however, H_2S itself does not cause oxygen deficiency.

11. D

12. B, false. This is an old H_2S myth.

13. B, false. This is another old H_2S myth.

14. C (most important), E, D, and B. Safe work practices (choice C) would prevent the truck driver, the facility operators, or anyone else from being in the vicinity of a potential H_2S plume. Choice E could have saved the lives of two of the three men. Choice A will not help— since the steps are adjacent to the silo, an alarm at the time of spill-over still means personnel can be trapped and exposed to the gas.

15. B. When the foreman is working at the valves, then he is closer to the evacuation window (through which the H_2S enters) than he is to the door. Even if the alarm goes off, he will not have time to escape.

16. B. The machinery area, where the foreman was working, does not meet the criteria for a confined space.

17. B. Confined space training is necessary for all operative employees. However, it probably would not have helped here, since the machinery area is not a confined space, and coworkers were not involved.

18. C. The air in the underground tank must be kept separate from areas occupied by people. D is not quite so satisfactory, because a potentially hazardous situation can arise during power outages, maintenance on the ventilation system, or malfunctioning equipment.

19. B, false. Biological methane can be produced in collection systems, if the circumstances are favorable.

20. B, false. Biological methane is generally produced under anaerobic circumstances. However, if there may be geological methane present, then whether the circumstances are aerobic or anaerobic is irrelevant.

21. B, false. At normal pressure and temperature, methane is not soluble in water. However, under high pressure (such as the pressures deep within the rock strata), methane is water soluble.

22. A, true. LELs (and LFLs) are not hard-and-fast boundaries defining absolute security. LELs can change depending upon factors such as combinations of gases present.

23. B, false. A methane concentration above the UEL cannot be taken as "safe" because there is always the possibility of dilution to an explosive level.

24. A. The engine will also produce CO_2 and use up oxygen, but because small amounts of CO can be fatal, carbon monoxide is the primary danger.

25. C. Carbon monoxide migrating through soil after the nearby use of explosives has been known to cause CO poisoning during the construction work on municipal sewer systems. Carbon dioxide is toxic at high levels—concentrations greater than 10%.

26. B. (And of course it is always necessary to take precautions against gasoline spills, regardless of where the engine is located.)

27. B. The gasoline-powered pump was installed in a confined space with almost no ventilation. The carbon monoxide generated by the engine had nowhere to go.

28. C. Either of these could have avoided the accumulation of CO in the sewer. If the gasoline-powered pump is placed outside the sewer, then it is important to ensure that the exhaust gases do not enter the sewer.

29. E. The confined space should not have been entered without safety checks, including testing the atmosphere. Rescuers who are not trained and equipped for the rescue attempt often fall victim themselves.

30. B, false. HIV is not believed to survive in collection systems.

31. E. There are no vaccinations against protozoa and worms.

32. F. The hepatitis diseases are separate diseases. Having had one does not confer immunity against another type.

33. B. There is no vaccine for hepatitis D or hepatitis C. However, hepatitis D requires hepatitis B to replicate and cause infection, so the hepatitis B vaccine can help protect against hepatitis D.

34. B, false. The bacterium that causes the disease, *Clostridium tetani*, is commonly found in sewer systems. It can enter through deep wounds or puncture wounds.

35. B, false. It is true that *Helicobacter pylori* has been classified as a Class I human carcinogen. However, studies have not found an increased prevalence of peptic ulcer among sewage workers.

References

1. Nogué, S., Pou, R., Fernández, J., and Sanz-Gallén, P. (2011). Fatal hydrogen sulphide poisoning in unconfined spaces. *Occupational Medicine*, 61(3), 212–214.
2. National Institute for Occupational Safety and Health (NIOSH) (1984). Fatal accident summary report: Two confined space fatalities during construction of a sewer line. Fatal accident circumstances and epidemiology database, incident FACE 8413. NIOSH, U.S. Department of Health, Education and Welfare, Washington, DC. Accessed January 14, 2016.

Index

A

Abbeystead explosion
  accident, root cause, 123–124
  background, 111
  continuing duty to advise, 128–129
  dead end, 114
  engineering aftermath, 124–127
  events, May 1984, 116–117
  findings, 125–126
  gas analysis, 125
  gases in tunnel, 119–120
  gas transfer to valve house, 121
  geological information and studies, 124–125, 129–130
  headaches, 132
  Health and Safety Executive investigation, 117–122
  "hic est dracones," 133
  historical mining, 131–132
  ignition source, 121
  IGS viewpoint, 131
  inquiry commissions value, 133
  introduction, 111
  legal aftermath, 128–129
  lessons learned, 132–134
  methane, 120, 129–133
  modifications selected, 126–127
  oil and gas company borehole, 129, 130
  open washout valve, 122–123
  outfall chambers, 114–116
  physical effects during tunneling, 132
  quantifying methane ingress, 124–126
  recommendations, 121–122
  redesign criteria, 126
  reservoir identification, 124–126
  simulation, 119–120
  sources of methane, 120
  transfer scheme, 111–116
  valve house, 114–116
  ventilation, 114, 132–133
  void in the tunnel, 118–119
  water analysis, 125
  water solubility, 133
  Wyresdale tunnel, 113
abdominal colic, 53
access control, risk reduction, 12
accidents
  blood values, 79–83, 84–85
  cause of death establishment, 76
  domestic poisoning, faulty drains, 95
  municipal sewage pumping station case, 94–95
  root cause, methane case, 123–124
  silo, water purification sludge, 95
  urine biomarker data, 79–83
  variations within same accident, 76
Accra, Ghana, 174
ACGIH, *see* American Conference of Governmental Industrial Hygienists
acquired immunodeficiency syndrome (AIDS), 185, 189, 191–192
activation of machinery, dangers, 11
adenosine triphosphate (ATP), 45–46, 50
aerosol shock, 186
AGA, *see* American Gas Association
age impact, 52
Agency for Toxic Substances and Disease Registry (ATSDR), 52, 165
AIDS, *see* Acquired immunodeficiency syndrome
airways reactivity, pulmonary edema, 51
alveolar macrophages (AM) impairment, 59–61, 63
AM, *see* Alveolar macrophage
American Conference of Governmental Industrial Hygienists (ACGIH)
  carbon dioxide, 171
  carbon monoxide, 166
  HCCPD concentrations, 176
  hydrogen sulfide, 26, 27
  overview, 3
American Gas Association (AGA), 144, 148, 155

Milton Keynes UK
Ingram Content Group UK Ltd.
UKHW040444071024
449327UK00020B/984